ROBOTIC GRASPING AND
FINE MANIPULATION

**THE KLUWER INTERNATIONAL SERIES
IN ENGINEERING AND COMPUTER SCIENCE**

ROBOTICS: VISION, MANIPULATION AND SENSORS

Consulting Editor

Takeo Kanade

ROBOTIC GRASPING AND FINE MANIPULATION

Mark R. Cutkosky
Stanford University

KLUWER ACADEMIC PUBLISHERS
Boston/Dordrecht/Lancaster

Distributors for North America:
Kluwer Academic Publishers
190 Old Derby Street
Hingham, MA 02043

Distributors outside North America:
Kluwer Academic Publishers Group
Distribution Centre
P.O. Box 322
3300 AH Dordrecht
The Netherlands

Library of Congress Cataloging in Publication Data

Cutkosky, Mark R.
  Robotic grasping and fine manipulation.

  (Kluwer international series in engineering and
computer science ;   )
  Bibliography: p.
  Includes index.
  1. Robotics.   2. Manipulators (Mechanism)   I. Title.
II. Series.
TJ211.C87   1985   670.42 '7   85-12723
ISBN-13: 978-1-4684-6893-9   e-ISBN-13: 978-1-4684-6891-5
DOI: 10.1007/978-1-4684-6891-5

TO PAMELA

CONTENTS

List of Figures   ix

List of Tables   xi

Preface   xiii

Acknowledgement   xv

1. **Introduction**   1
2. **Previous Investigations of Fine Manipulation and Grasping**   5
   2.1 Fine Motion and Control   5
   2.2 Robotic Wrists   8
   2.3 Applications to Assembly and Surface Finishing Tasks   10
       2.3.1 Assembly   10
       2.3.2 Surface Finishing   11
   2.4 Passive Hands and Grippers   11
   2.5 Active Hands and Grasping   12
3. **Robot Tasks in a Metal-Working Cell**   17
   3.1 Task Descriptions   17
       3.1.1 Materials Handling   17
       3.1.2 Assembly   18
       3.1.3 Grasping   23
       3.1.4 Surface Finishing and Shaping   23
             3.1.4.1 Contour following   26
             3.1.4.2 Contour modification   28
   3.2 Discussion: Coupled Fine and Gross Motions   30
4. **A Wrist for Fine-Motion Tasks**   32
   4.1 Wrist Description   32
   4.2 Control Architecture   37
   4.3 Experiments   39
       4.3.1 Contour Following   40
             4.3.1.1 State estimation   42
             4.3.1.2 Control law   46
       4.3.2 Grinding   51
   4.4 Discussion of Results   53
5. **Analysis for an Active Robot Hand**   55
   5.1 The Promise of Further Dexterity   55
   5.2 Introduction to Grasp Analysis   56
       5.2.1 Grasping Model and Assumptions   58
       5.2.2 Stiffness, Strength and Stability of a Grasp   61
       5.2.3 Procedure for Establishing Grip Properties   62
       5.2.4 Two-Dimensional Examples   63
             5.2.4.1 Choosing among five grips: an example   65
             5.2.4.2 An unstable example   69
   5.3 Extension to Three-Dimensional Problems   74
       5.3.1 Forward Force and Displacement Relations   74
       5.3.2 Summary of Forward Transformations   77

5.3.3 Finger Motions and Constraints	78
5.3.4 Constraints at a Contact	79
5.3.4.1 Case 1: exactly determined	80
5.3.4.2 Case 2: under determined	81
5.3.4.3 Case 3: over determined	82
5.3.5 Computing Changes in Grip Force	83
5.4 A Closer Look at Contact Conditions	87
5.4.1 Point Contact	90
5.4.2 Curved Finger Contact	91
5.4.2.1 Effects of rolling motion	93
5.4.3 Very Soft Finger	95
5.4.3.1 Effects of deforming fingertip	103
5.4.4 Soft, Curved Fingertips	105
5.5 Examples	109
5.5.1 Pointed Fingers	109
5.5.2 Procedure for Left Finger	111
5.5.2.1 Discussion	113
5.5.3 Curved Fingertips	113
5.5.3.1 Discussion	115
5.5.4 Very Soft Fingers	116
5.5.4.1 Discussion	117
5.6 Summary	118
6. Natural Examples of Grasping	**123**
6.1 The Human Hand	123
6.1.1 Conformability	124
6.1.2 Muscles	125
6.1.3 Hand/Wrist Interaction	125
6.1.4 Finger Coupling and Specialization	126
6.1.5 Grasps	126
6.1.6 Sensation and Control	132
6.2 Other Natural Examples	136
7. Designing Hands and Wrists for Manufacturing	**140**
7.1 Wrist Design	141
7.2 Hand Design	143
7.2.1 Grasping *vs* Manipulation	143
7.3 Control	150
8. Summary and Conclusions	**153**
Appendix for Grasp Analysis	**157**
A.1 Matrix Identities	157
A.2 Matrix Method for Under Determined Finger Motions	158
A.3 Differential Jacobians	159
A.4 Rolling Contact	160
A.5 Details for Examples in Section 5.5	161
References	**165**
Index	**175**

LIST OF FIGURES

Figure 3-1: Materials handling with large 6-axis manipulator (from [97])   18
Figure 3-2: Materials handling with manipulator and track (from [97])   19
Figure 3-3: Beginning of insertion of peg in hole   20
Figure 3-4: Insertion with and without decoupled force-deflection behavior   21
Figure 3-5: Properties of surface finishing tasks in a metal-working cell   24
Figure 3-6: Tracking a contour with misalignment and roughness (from [29])   27
Figure 4-1: Design of Compliant Wrist (adapted from [103])   34
Figure 4-2: RCC Linkage (from [28])   35
Figure 4-3: Hydraulic Schematic of Wrist (from [29])   35
Figure 4-4: Communications for robot sensory control (modified from [29])   36
Figure 4-5: The instrumented wrist inserting a peg into a hole   39
Figure 4-6: Variable-step tracking of a contour with an unexpected corner   40
Figure 4-7: Implementation of Kalman filter estimator   44
Figure 4-8: Least-Squares Prediction of Parametric Trajectory   46
Figure 4-9: Tracking a misaligned straight contour -- no predictor   50
Figure 4-10: Tracking a misaligned straight contour -- two point predictor   50
Figure 4-11: Tracking a misaligned straight contour -- ten point predictor   51
Figure 5-1: A two dimensional object held by three fingers   60
Figure 5-2: Detail of a single two-dimensional finger from Figure 5-1   62
Figure 5-3: Five ways to grip a rectangle with four fingers   66
Figure 5-4: Grip stiffness for force at angle θ_q 67
Figure 5-5: Maximum force without slipping at angle θ_q 68
Figure 5-6: Instability of a rectangle held by two fingers   70
Figure 5-7: Coordinate systems for a finger touching an object   75
Figure 5-8: Flow chart for forward force and displacement transformations   78
Figure 5-9: Flow chart for cases in which [P] is invertible (Case 1)   82
Figure 5-10: Flow chart for relationships between displacements and forces (Case   83
             2 or 3)
Figure 5-11: (from Salisbury [19])   88
Figure 5-12: Examples of fingertip geometry   89
Figure 5-13: Rolling contact   90
Figure 5-14: Cross section of a large-radius hemispherical fingertip on a flat object   94
             surface
Figure 5-15: Cross section of a small-radius hemispherical fingertip on a flat object   95
             surface
Figure 5-16: Elastic fingertip in contact with object surface (perspective)   96
Figure 5-17: Elastic fingertip in contact with object surface (section)   97
Figure 5-18: Pressure distributions for elastic, soft, and very soft fingertips 106
Figure 5-19: Maximum shear stress for moment about finger axis   108

Figure 5-20: Holding a rectangle between two fingers — 3 examples   110
Figure 5-21: Curved fingertip   114
Figure 5-22: Curved fingertip after rolling $\delta\theta$ with respect to object   115
Figure 5-23: Relations between finger models   121
Figure 6-1:   Power grasp, with fingers and thumb curled about a heavy object   127
Figure 6-2:   Cylindrical grasp for a heavy, convex workpiece   128
Figure 6-3:   Five-fingertip grasp for light cylindrical objects   129
Figure 6-4:   Modified hook grasp with thumb along tool   130
Figure 6-5:   Modified cylindrical grasp with partial fingertip prehension   131
Figure 6-6:   Modified precision grasp for objects that are not small with respect to 132
              the hand
Figure 6-7:   Three-finger precision grip using thumb, index and middle fingers   133
Figure 6-8:   Two-finger precision grip, or "palmar grip"   134
Figure 6-9:   Lateral pinch grip   135
Figure 6-10: A complex, two-handed grip   136
Figure 6-11: Some of the sensors in the skin of a hand (from [130], reprinted with 137
              permission from W. B. Saunders Co.)
Figure 6-12: Automatic grasping responses in infants (from [134], reprinted with 138
              permission from The Pergamon Press Ltd.)
Figure 6-13: Typical arrangement of muscle receptors (from [133], reprinted with 139
              permission from The Scientific American)
Figure 7-1:   Roll, pitch and yaw manipulations with the fingers   144
Figure 7-2:   Three-fingered hand showing wrap, and two- and three-fingered pinch 146
              grasps
Figure 7-3:   Simplified, anthropomorphic two-fingered hand   147
Figure 7-4:   Communications scheme for arm-wrist-hand combination   151
Figure A-1:   Matrices for left finger - pointed or rolling contact   162
Figure A-2:   Matrices for left finger - pointed or rolling contact   163
Figure A-3:   Compliance matrix for soft finger example   164

LIST OF TABLES

Table 5-1: Soft fingertip deflections for 4.0 N load and 1 cm^2 or 4 cm^2 contact area   104

Table 5-2: Motions of left and right finger and object contact areas (pointed fingertips)   111

Table 5-3: Contribution from left finger to $\delta g_{b'}$ (pointed fingertip)   112

Table 5-4: Change in g_b due to motions dx, dy, and $d\theta z$ (pointed fingertips)   113

Table 5-5: Contribution from left finger to δg_b (rounded fingertip)   116

Table 5-6: Change in g_b due to motions dx, dy, and $d\theta z$ (rounded fingertips)   116

Table 5-7: Change in δg_b for small contact area (soft fingertips)   117

Table 5-8: Change in δg_b for large contact area (soft fingertips)   118

Table 5-9: Summary of contact models derived in Section 5.4   120

Preface

When a person picks up a metal part and clamps it in the chuck of a lathe, he begins with his arm, proceeds with his wrist and finishes with his fingers. The arm brings the part near the chuck. The wrist positions the part, giving it the proper orientation to slide in. After the part is inserted, the wrist and fingers make tiny corrections to ensure that it is correctly seated. Today's robot attempting the same operations is at a grave disadvantage if it has to make all motions with the arm. The following work investigates the use of robotic wrists and hands to help industrial robots perform the fine motions needed in a metal working cell.

Chapters 1 and 2 are an introduction to the field and a review of previous investigations on related subjects. Little work has been done on grasping and fine manipulation with a robot hand or wrist, but the related subjects of robot arm dynamics and control have an extensive literature.

Chapter 3 takes a closer look at several tasks in a metal working cell including assembling, polishing, grinding and machining parts. The tasks are broken into subtasks distinguished by

- Large or small motions. (In the example above, the subtask performed by the arm involves large motions, and that of the wrist involves small motions.)

- The need to adapt to an existing geometry (as in polishing a part) or to establish a new one (as in grinding to a prescribed line).

- Dependence or independence of time and velocity. (In grinding, a smooth finish requires control of the velocity along the part, but in assembly, pauses can be tolerated.)

A single manipulator is unable to fulfill the diverse task requirements. Today's large industrial robots are designed for large motions and high speeds, and for establishing a geometry or trajectory in space. However, if a second system is added for small and slow motions, and for adaptation to constraining geometries, most of the selected tasks can be accomplished. For some tasks, such as assembly, the gross motion and fine motion systems (in the present study, the robot and the wrist) can function independently. More difficult are tasks, such as disk grinding, for which the fine and large motions are coupled. For such tasks it is necessary to coordinate the actions of the arm and wrist.

Chapter 4 describes the use of an instrumented, controllable wrist with a large industrial robot. The wrist permits fine motions that require compliance or adaptation to an existing geometry. Experiments focus on tasks that require communication between the robot arm and wrist. The results suggest that it is still appropriate to treat the wrist and robot separately and to take advantage of the simplicity that a distributed control system affords.

For many manufacturing tasks an arm and wrist provide sufficient dexterity, but for more delicate manufacturing tasks a hand with actively controlled fingers may be needed. An active hand permits the robot to attempt the tiny movements that people make in threading a bolt into a tapped hole or when using a small screwdriver to turn a screw. With sensors in intimate proximity to the object and the task, the hand can respond rapidly to changes in loading and adjust the gripping force and rigidity as required.

Chapter 5 begins with a discussion of the formidable problems that must be resolved before active robot hands can be put to use in the manufacturing environment. A first step is to construct a mathematical model of an active gripper handling a tool or a workpiece. A method is presented for determining mechanical properties of a grasp and predicting how the grasp will behave in response to task-induced forces and motions. On this basis, competing gripper designs may be compared or, for a single gripper, competing grasps may be compared. The results depend strongly on the contact conditions between the gripper and the object and particular attention is given to the effects of pointed, curved, soft and hard contacts.

The analytic approach is complemented with a more introspective look at our own wrists and hands, and those of other creatures. Chapter 6 describes how nature reveals inspirational models of distributed control, of communication and coordination between arms, wrists and fingers and of partitioning for fine and gross motions. The human hand has been a model for several recently developed research grippers. Attributes of human hands, and those of simpler animals, are considered for the design and control of a hand for manufacturing. The human hand is an excellent source of ideas, but it has evolved for a much wider world than a flexible manufacturing system represents and is a more complex organ than manufacturing tasks require.

The investigation of grasping and fine manipulation reveals general principles concerning arm-wrist-hand systems for manufacturing tasks. Chapter 7 summarizes these principles and presents guidelines for the design of active wrists and hands.

Acknowledgement

This book is adapted from a doctoral thesis, "Grasping and Fine Manipulation for Automated Manufacturing," submitted to the Department of Mechanical Engineering at Carnegie-Mellon University in January 1985. The book is the result of several pleasant years within the Robotics Institute and the Mechanical Engineering Department at Carnegie-Mellon University. Thanks are due to Professors Dwight Baumann, Matt Mason, Mark Nagurka and Raj Reddy for their advice and support. Above all, I thank Professor Paul Wright for his constant guidance and encouragement in this project and for numerous inspirational conversations on why we hold things the way we do. I also thank machinists Jim Dillinger and Dan McKeel for their expert opinions on grasping in a manufacturing environment and for the use of their hands in Chapter 6. The work leading to this book was supported by a fellowship from the Phillip J. McKenna Foundation with additional support from the Robotics Institute.

CHAPTER 1
Introduction

Like animals, machines interact with their environment using grippers. In animals, grippers take a wide array of forms including mouths, pincers, hands and tentacles, all designed for a set of grasping and manipulation tasks. Many animals use their grippers not only as organs of action, but also as exploratory organs, relaying information about the outside world. The gripper becomes a bridge between the animal and the world around it. In manufacturing, grippers traditionally have taken the form of tongs or clamps. Like animal grippers, they can hold a part securely, but unlike animal grippers, they cannot manipulate objects and cannot provide information.

With the advent of industrial robots, grasping and manipulation have assumed a new importance. The primary purpose of industrial robots is to grasp objects and to manipulate them in fulfillment of a task, such as fitting parts together or loading them into a machine. The first industrial robots were simple devices relegated to repetitive tasks with a minimum of precision or flexibility. The objects these robots moved were simple shapes, such as cylinders of metal bar stock, and the grippers were two-fingered devices resembling pliers or tongs.

As the reliability and precision of industrial robots have increased, robots have been applied to a variety of unusual tasks. Many of these applications involve dangerous or unpleasant working conditions, exacting quality control requirements or other features that make it especially attractive to use robots in place of human workers. In a carefully defined task, the robot can be designed to perform better than a person. In an automated forging system, huge manipulators may transport billets weighing hundreds of pounds while, at the other end of the spectrum, tiny robotic arms can place electronic parts weighing milligrams, on circuit boards. Special purpose grippers allow robots to grasp hot forgings, glass plates, automobile tires and pieces of cloth. The grippers employ everything from rubber fingers to a vacuum to securely hold their workpieces. Most industrial grippers are designed on an ad hoc basis, to perform a specific task with specific parts. They may be precise and reliable, but they are even less flexible than the general purpose two-fingered devices used in early applications. A gripper designed with suction cups to handle panes of glass is not suitable for handling machined parts and a magnetic gripper for steel parts becomes useless for picking up aluminum.

The current trend is to apply robots to flexible manufacturing operations: systems that can produce small batches of different parts without re-tooling. A number of investigators have proposed systems composed of independent "cells." Each cell is dedicated to a specific process such as forging, machining or welding. Robots are responsible for manipulating and assembling parts within the cell. The strongest advantage to cells is their ability to switch between the variants of a basic part style simply by loading new parts programs into the machines of the cell. For example, computer-controlled machine tools can machine parts ranging in size from an inch to over a foot on each side. The shape of the parts is limited only by the sophistication of the parts programs and the number of axes on the machine tool. Most of the hardware required for flexible cells is available. This includes robots, computer-controlled processing machines, sensors and programmable measurement devices. However, a number of difficulties prevent flexible manufacturing systems from functioning without human assistance. Foremost are the difficulties in accurately grasping, manipulating, finishing and assembling parts. The sources of these difficulties are two-fold:

- General purpose industrial manipulators cannot achieve both large motions and fine-motion manipulation in flexible manufacturing systems.

- The current generation of manufacturing grippers and fixtures cannot grasp a wide range of part shapes.

The first problem stems from the design of industrial robots, which represents a compromise among accuracy, speed, payload and flexibility. As a result, robots cannot accurately load heavy workpieces into fixtures or assemble them. Even with sensory feedback, large robots lack the resolution and responsiveness for fine motions. The responsiveness of industrial robots is likely to improve, but a better solution is to use an end-effector for fine accommodations. This is the solution that people use: The arm provides the approximate position and orientation while the hand and wrist perform small accommodations. Unfortunately, the current generation of industrial end-effectors cannot manipulate parts.

The second source of difficulty results from the special-purpose nature of current industrial grippers. This has come to the attention of several investigators working with manufacturing systems. W. Gevarter writes:

> "Nevertheless, though hundreds of different special-purpose end-effectors now exist, the end-effector remains one of the major limiting factors in universal robot manipulation due to the lack of dexterity and programmability of the hands." [1]

One solution is to provide robots with an array of grippers or gripper-adaptors. This is practical in many instances but it suffers from a few drawbacks: It becomes expensive to keep an inventory of grippers, the robot slows down if it pauses frequently to change grippers, and if tactile and force sensors are added, it becomes difficult to make connections between the gripper and the wrist.

An alternate solution is to construct more versatile grippers. A universal gripper would be able to grasp every part the robot might encounter. Recent universal gripper designs have used the human hand as a source of inspiration. In fact, some of the most versatile grippers are prosthetic devices. The human hand, however, is many things in addition to being a gripper. It is a sensory organ designed as much for palpation as for manipulation. The hand is also an extremely general-purpose end effector, at least as well suited to kneading dough and playing the violin as it is to working in a factory. In fact, the hand is only moderately well suited to manufacturing tasks. When a mechanic starts to work on a machine the first thing he reaches for is his toolbox. This suggests that by studying the mechanics of grasping and manipulation for a set of tasks it should be possible to specify a gripper that exceeds the performance of the human hand for the same tasks. In any event, an understanding of the mechanics of grasping and manipulation for a set of tasks makes it possible to evaluate various gripper designs and to specify control laws for them.

An understanding of grasping also paves the way for robots that know how to grasp parts. In today's manufacturing cells, the instructions concerning how to pick up a part are stored as part of the information associated with the part. It is the programmer and not the robot who decides the ideal grasp. However, as robots become more sophisticated they will be put to work in places where they must function in a more autonomous fashion. Robots harvesting the ocean floor or working in a nuclear power plant will be much more effective if they can determine how to grasp and manipulate the objects they encounter. This requires quantifying the suitability of a grip. Usually, the geometry of a specific object and task will limit the choice of grips. Competing grips may then be compared in terms of strength, stability and stiffness. The same criteria can be used to compare gripper designs.

To design and control a truly general-purpose arm and end effector for grasping and manipulating objects is a formidable task. This book proposes a framework in which the design and control become manageable. In particular, several limitations and assumptions are used to simplify the problem:

- The investigation of gripping and manipulation focuses on robot tasks in an automated metal-working cell. However, the analyses can be extended to most manufacturing environments.

- Once the basic arm and grip geometries have been established, only small motions are made with the wrist and hand.

- In the hierarchy used, the arm, wrist and gripper function as modules, each performing a subset of the task, but communicating among themselves and coordinating their actions where necessary.

CHAPTER 2
Previous Investigations
of Fine Manipulation and Grasping

2.1 Fine Motion and Control

Efforts to improve the performance of industrial robots in fine manipulation tasks have employed both active methods, in which the robot controller compensates for sources of error, and passive techniques, involving mechanical redesign of the arm.

Much of the inaccuracy of industrial manipulators can be traced to an imperfect model of the robot kinematics and to deflections caused by gravity, task related forces and inertial terms. For fine manipulations, the inertial terms may be ignored, but gravity and task-induced forces combine with compliance and backlash in the arm to make the manipulator imprecise. A number of methods may be used to improve the accuracy of manipulators. Efficient computational schemes [2-6] permit dynamic and gravity-loading terms to be continuously computed and fed forward in the control system, resulting in a substantial improvement in accuracy and stability. Such feed-forward methods are beginning to be applied to commercial robots [7]. Strains within the arm or drive train may be measured and used in active control schemes [8-11], incorporating a model of the elastic flexibility of the arm, to make the tip of the robot dynamically stiff. Adaptive control methods [12-15] promise accurate control without explicit modeling of gravity and dynamic terms or of friction and elasticity in the drive train.

When the above techniques are applied, the remaining robot inaccuracies are due largely to imperfect modeling of the robot kinematics. In an effort to develop robots that are inherently more accurate, Asada [16, 17] proposes direct-drive arms in which each manipulator joint is connected directly to a rare-earth DC torque motor. The direct drive mechanism eliminates the usual power transmission problems of industrial robots in which backlash, "stiction" and drive train compliance prevent fine control of the arm.

Constraint and compliance

The above methods promise substantial improvements in the positional accuracy of industrial robots, but even when they are adopted, it remains useful to match the compliant characteristics of the robot to those of the task. One reason is that no matter how accurately the arm itself is controlled, there will always be minor variations in the external working environment. For example, the workpiece may be slightly misplaced and its surface may reveal unexpected bumps or rough spots. Mason [18] shows that in many robot tasks, a set of cartesian axes can be found such that the task constrains the robot in certain directions but not in others. For example, in the surface finishing operations discussed in the next chapter (in Section 3.1.4), the robot is constrained by the surface in the direction perpendicular to the surface, but can move in the plane of the surface. Similarly, when pushing a peg into a hole as described in Section 3.1.2, the robot is constrained in the radial direction but unconstrained along the axis of the hole. The directions in which the robot is *unconstrained* are candidates for position and velocity control; the robot is free to control motion in those directions. In the *constrained* directions, the robot has the ability to exert forces and, therefore, the constrained directions are candidates for force control. To partition a task into axes of motion control and force control is to find a *C-surface*, or "constraint surface" for the task [18]. When the robot is force-controlled in the constrained directions, it can be guided by the constraints of the task, without allowing excessive contact forces to develop. As a result, it becomes unnecessary to establish the exact position and orientation of the workpiece or to program the robot precisely.

Several techniques have been suggested for controlling the motion of the robot in certain directions while controlling force in others. Often, the forces are sensed with an instrumented wrist at the tip of the robot [19-29] . If the robot has good force reflecting characteristics (as in the case of a direct-drive arm) the wrist may be eliminated and forces at the hand may be determined by measuring the torques at the manipulator joints [30]. Once the cartesian directions of force and motion control have been identified, a "figure of merit" may be used to determine which manipulator joints are best suited for force servoing or position servoing [31, 32]. Inaccuracies that result from this method may be reduced with a differential correction technique [22]. More recent efforts to simultaneously control force in certain directions while controlling motion in others include "Hybrid Position/Force Control" [33], "Resolved Motion Force Control" [34], velocity modifications based on measured forces at the hand [20, 21] and similar approaches in which force information is used to modify a position feedback loop [35].

The *C-surface* concept may be extended to cover cases for which motion is unconstrained, but subject to friction, or for which motion *is* constrained, but non-zero [18]. However, there are other tasks for which the constraints cannot be identified simply as force constraints or motion constraints. For example, suppose the robot follows a surface which is soft. The robot is no longer completely constrained in the direction perpendicular to the surface. Similarly, if the robot pushes a hollow plastic peg into a hole, it is no longer strictly confined to motions along the axis of the hole. In such cases, the compliance in the constraints makes it convenient to think in terms of controlling the stiffness or compliance of the manipulator [19, 36, 37]. The robot should be compliant where the environment is stiff and vice-versa. Perfect force control or position control can be seen as extreme examples of compliance and stiffness, respectively. As the robot becomes perfectly compliant, changes in position have no effect on the force exerted by the robot. The force applied by the robot resembles the pull of gravity. A similar effect can be achieved if the robot is controlled, with an accurate force-servo system, to maintain a constant force. At the opposite extreme, if the robot is stiff, small changes in position produce large changes in the robot force. For a position-controlled robot, the stiffness of the robot will be a function of the feedback gains in the position servo loops. Salisbury [19] explores this relationship to specify the stiffness of a manipulator in cartesian coordinates. The stiffnesses of the manipulator joints are related to the stiffness of the robot hand through the Jacobian matrix of the manipulator. Thus, for the peg-in-hole assembly problem discussed in Section 3.1.2, the servo gains would set so that the stiffness matrix of the robot/gripper structure becomes diagonal toward the tip of the peg.

At low speeds, stiffness and compliance represent the mechanical impedance and admittance of a system. Hogan [36, 37] suggests that the stiffness control and position/force control may be viewed as special examples of impedance control. The interaction between the robot and its environment is a problem in minimizing deviations from desired motions while simultaneously minimizing interaction forces. The solution is to match the robot impedance to the environment admittance.

2.2 Robotic Wrists

Any of the above techniques will help a robot to assemble parts or to track a contoured surface in space and many of them have been demonstrated in experiments. However, all other things being equal, they work best when used with a responsive, high-bandwidth manipulator. To use a 2000lb manipulator arm with a 3Hz bandwidth for assembly or contour tracking is a bit like trying to thread a needle or to write with a pencil using only the muscles in our arms and shoulders. When we use our own arms and hands in assembly, we perform fine manipulations with the fingers and wrist and leave gross motions to the arm. Perhaps inspired by this natural partitioning of the task, a number of investigators [26, 28, 38, 39, 40, 41] have suggested wrist devices with active or passive compliance, mounted at the end of the manipulator arm. The wrists add a certain amount of complexity, but enable large manipulators to be used for fine-motion tasks from which their size and lack of precision would otherwise preclude them.

Wrists may demonstrate active control, passive compliance or a combination of the two. For assembling parts, Nevins *et al* [42, 43] have described a passive *remote-center-compliance* (RCC) wrist, to be sandwiched between a gripper and the end of a robot arm. The wrist is designed so that its stiffness matrix is diagonal at a point located some distance out from the center of the wrist. As discussed in Section 3.1.2, for parts mating it is ideal if the stiffness matrix of the part and the structure supporting it becomes diagonal at the tip of the part. In other words, it is ideal if forces at the tip of the peg cause the peg to translate without tilting, and moments at the tip of the peg produce a pure rotation. The RCC wrist achieves this effect if the distance between center of compliance and the wrist is approximately equal to the combined length of the gripper and workpiece. Other RCC devices are described in [44]. RCC devices have found successful application in industry, and commercial versions are now sold. No sensory information or active control is needed to use an RCC wrist for peg-in-hole assembly tasks; the passive compliance of the wrist provides the necessary fine accommodations independent of the robot. In a sense, the RCC constitutes a mechanical implementation of a fine motion control scheme that is suited to assembly tasks. The only serious drawback to passive RCC devices is that they are perfectly suited only for gripper/part combinations of a single length.

To extend the flexibility of the passive RCC, and to explore its use in fine manipulation tasks other than assembly (contour following, for example) instrumented RCC devices have been developed [28, 42]. Deflections within the wrist are measured and used to modify the robot path. Since the wrist is compliant, deflections are proportional to the forces applied to the wrist. The RCC effect remains useful since it often simplifies the control algorithms by decoupling forces and deflections at the tip of the tool or workpiece.

Further versatility is possible with a servo-controlled wrist. Van Brussel *et al* [39, 40] discuss wrists with compliant electric and hydraulic servo drives. Forces are measured and used to control the position of the wrist. The control objective is essentially

$$\Delta x = C \, \Delta f$$

where C can be viewed as a compliance matrix for the device. If C is diagonal, a diagonal stiffness matrix for the device results. Setting C to zero results in a perfectly stiff (within the limitations of the hardware) position-controlled wrist. Sharon and Hardt [41] discuss a similar hydraulic servo-controlled wrist with 5 degrees of freedom (three translations and two bending rotations). They consider the problem of dynamic interaction between the wrist and the robot arm and show that coupling can be ignored if the bandwidth for the wrist/robot control loop is kept below the fundamental vibrational frequency of the robot. At higher frequencies, active control methods, such as those mentioned above in Section 2.1 and discussed at length in [8, 9, 10, 45], must be employed to keep the system stable.

2.3 Applications to Assembly and Surface Finishing Tasks

2.3.1 Assembly

Robot assembly has been an active area of investigation for several years. Assembly efforts have included inserting electronic components, such as transistors, on circuit boards [46], placing armatures, bushings and end housings on motors [26], pressing bearings on shafts and inserting valves in cylinders [47]. Many assembly tasks, including the ones listed above, involve circular symmetry — which reduces the task from a three-dimensional problem to a two-dimensional one. Theoretical investigations of assembly have focused on the idealized model of inserting a cylindrical peg into a cylindrical hole. In fact, inserting a peg into a round hole is possibly the single most frequently performed task in robotics research. The mechanical equations have been derived for the case in which the peg and hole are made of rigid materials [42, 44, 48] and for selected cases in which either the peg or hole is deformable [38, 49]. The analysis has also been extended to the assembly of rigid pieces containing multiple pegs and holes [50] and to interference fits [51]. For a rigid peg and hole of given dimensions, the two most important variables are the angular misalignment of the peg and the coefficient of friction. Once the tip of the peg makes contact with the chamfer surrounding the hole, the assembly will continue smoothly as long as the angle between the axes of the peg and the hole does not become sufficiently large for jamming or wedging to occur [44]. If jamming or wedging *does* occur, pushing harder will not help; the peg must be straightened and/or the coefficient of friction reduced through lubrication. The coefficient of friction is beyond the control of the robot, but a number of methods may be used to ensure that the angular misalignment of the peg is kept to a minimum.

2.3.2 Surface Finishing

Robot finishing operations including grinding, routing and deburring recently have become areas of research and development [47,52-57] . In most cases, the robots are used exclusively for grinding or deburring. This makes it possible to choose robots that are slower, stiffer and more accurate than the general-purpose manipulators used in. flexible manufacturing cells. Hirzinger [47] uses an instrumented wrist unit mounted between the end of the robot arm and the grinding tool to measure forces and torques during the deburring of iron castings. The robot is able to determine the position and orientation of the casting by touching it in selected locations and to compensate for reductions in the diameter of the grinding wheel due to wear. The robot moves along the casting exerting a constant force against the grinding wheel so that it automatically slows down where large or hard burrs are present. Asada [17] describes a grinding tool guide for following the surface of an object. The guide employs two rollers and a grinding tool mounted on springs. The guide rolls along the object surface, ensuring that position errors between the grinding wheel and the object are small. The robot is required only to push the guide along, keeping it pressed against the surface. One could even imagine a self-propelled guide that only required the robot to place it upon the surface. The guide would crawl along the object surface, possibly held on by magnetic attraction, and remove material from the surface. The guide proposed by Asada is suited only to surface finishing tasks in which the grinder follows the existing object shape without significantly modifying it. The guide is limited also to contours that do not experience sharp changes in curvature.

2.4 Passive Hands and Grippers

A number of research grippers have been developed that can grasp a much wider array of part shapes and sizes than the grippers commonly used in industry. Most of these are passive, but a few have the ability to actively manipulate an object.

Passive gripper designs are surveyed and analyzed in [58, 59, 60]. Passive, flexible grippers may have fingers that can settle independently against a workpiece of arbitrary shape [61] or very flexible fingers that curl around objects [62, 63, 64]. Alternatively, passive grippers may use compliant materials or membranes filled with particles that conform to the object shape. The conformal designs often employ magnetic attraction [65, 66] or suction [67, 68] to hold the workpiece securely.

Some designs [69, 70] make a deliberate effort to achieve the different grasping modes that a human hand adopts when carrying a suitcase, turning a doorknob or inserting a key into a lock. Others [71] take a more abstract approach in which the gripping configuration can be manually reconfigured to suit a new part style.

Since most grippers are custom designs, their analysis is generally done on an individual basis. However, several recent articles [60, 72, 73] discuss general methods for determining the grasping and actuating forces in common gripper designs. In exacting applications, for example when handling fragile objects, it is useful to add sensors at the jaws or fingertips to measure task-induced forces and to permit control of the gripping force [46, 67, 74].

2.5 Active Hands and Grasping

A few active grippers have been developed, including three-fingered hands with jointed fingers [75-81] and non-anthropomorphic two-fingered devices with rotating belts [82, 83]. The analysis and control of active hands are formidable challenges, and the research in this area is in an infant stage. Hollerbach [84] discusses mechanical problems in hand design and control, observing:

"Mechanical impediments exist at all levels, including actuation transmission, materials and sensing. We need only think of the large, heavy, and slow manipulators of today to appreciate the scope of the problem in building a very small multi-manipulator, with three or four degrees of freedom for each of several fingers."

Recently, a few works have appeared that address grasping kinematics, gripper control and related topics.

Asada [85] begins by describing the force balance for an object held by a gripper with several fingers. He assumes that the gripper has k_a actuators each driving l fingers of which m are actually touching the held object at a particular time. Thus there are a total of $k_a \times m$ fingers in contact with the object, of which k_a are independent. He next assumes that each finger has a small contact area so that the contact between each finger and the object can be treated as a point contact. With this assumption the force exerted by each finger can be resolved into forces perpendicular and tangential to the object surface. The assumption is a limiting one because it removes the possibility that a single finger can apply a torque about its own axis and ignores rolling or rocking motion between the fingertip and the object. However, it is often a reasonable approximation for grippers with small gripping surfaces made of hard materials (a pair of tongs, for example). Generally, the point-contact approximation results in an overestimate of the gripping force required to maintain equilibrium. Salisbury [78] and Okada [75, 76] make similar assumptions in describing the forces exerted by their three-fingered hands, although Salisbury discusses the effects of having a "soft" finger that can apply moments, twisting about its central axis.

Having described the equilibrium requirements for an object held by several fingers, Asada addresses the problem of choosing a suitable finger configuration. He treats the held object as a rigid body and models the fingers as elastic members with one degree of freedom, along a specified trajectory or locus. He simplifies the grasping model by ignoring friction at the contact points between the fingers and the object. With this model he is able to construct a potential function, based on the shape of the object, which indicates the relative stability of different finger configurations. In the absence of friction, an object held in a stable grasp will return to its original position if displaced slightly. The theory works well for slippery objects and whenever the chief concern is that the object should not be dropped (when we wash dishes we hold them in a stable grasp). Unfortunately, the utility of the model for industrial robots is limited. Friction is an important consideration and is often used to advantage. According to Asada's theory there is no satisfactory way for a two-fingered gripper to grip many shapes. For example, there is no "stable" configuration for a two-fingered gripper grasping a sphere. In practice, people depend on friction when they design grippers and when they program robots to grasp and manipulate objects. A stable grip guarantees that the gripper will not drop an object, but often a great deal more is required. It may be required that none of the fingers of a gripper should slip with respect to an object while it is manipulated because if they do, the object will not return to the same equilibrium position.[1]

---

[1] In the absence of friction the object would return to its original, stable position.

Industrial robots often are programmed based on a precise knowledge of the position and orientation of a grasped object with respect to the robot coordinate system. As soon as any of the fingers slip, this information is lost.

Salisbury [77, 78] and Okada [75, 76] are concerned with developing control laws for multi-jointed three-finger grippers. The hand designed by Okada can perform a variety of manipulation tasks such as screwing a nut onto a bolt and manipulating a match box in three dimensions. When the motions of the manipulated object are not very small it becomes necessary to treat the fingertips not as points but as surfaces of finite radius. The fingertips roll with respect to the manipulated object and the kinematic description of the fingertip locus becomes extremely complicated.

Salisbury [77, 78] draws upon his earlier work in manipulator control [19] in which he discusses how to determine the correct servo stiffnesses for the joints of a robot to achieve some desired set of stiffnesses expressed with respect to the robot hand (or any other convenient coordinate system). Salisbury considers several contact types, including point, line and planar contact (with and without friction), and discusses the constraints imposed on the object by each. He also introduces a "soft" finger with friction which can apply torques about its own axis in addition to forces at the contact point. The effects of rolling and deformation of the fingertip are not considered. He also considers the interaction of groups of contacts about an object and discusses the conditions under which arbitrary motions and forces can be applied to the object. Like Asada, he shows that for pointed fingers a jacobian matrix can be found which relates forces exerted at the fingertips to an equivalent force and torque at the centroid of an object held by the fingers. He augments this jacobian matrix to include internal forces which essentially measure the magnitude of the "pinch" exerted by pairs of opposed fingers.

Hanafusa et al [86] also consider the kinematics of an object held by point-contact fingers. They discuss the conditions under which the object is free to move in any direction if the finger joints are loose, but becomes completely constrained if the finger joints are locked. They consider a gripper with an arbitrary number of fingers, each with an arbitrary number of joints, and discuss methods for specifying the redundant degrees of freedom.

Orin and Oh [87] have considered the related problem of determining the optimum distribution of forces in closed kinematic chains. They are primarily interested in extending earlier work in the control and analysis of walking machines, but they point out that a walking machine and a multi-fingered gripper handling an object both contain closed kinematic chains.[2] Usually, the number of independent joint actuators in the chains is greater than the number required to impart a desired set of velocities and forces to the body of the walking machine or the grasped object. They compute the dynamic terms for all the legs and then use a linear programming approach to minimize the energy expenditure in the motors, subject to a number of constraints. The constraints include friction limitations at the feet (or fingertips) and maximum joint torques at the actuators. Normally, friction limitations result in non-linear constraints, but by approximating the conventional "friction cones" with friction pyramids a conservative set of linear inequality equations is obtained. The contacts between the feet and the ground are treated as point contacts in the kinematic model. To measure the energy expenditure, a power term is established, where the power at each joint is a function of the joint torque and velocity. The simplex method is used to find the minimum of the cost function.

The linear programming method described by Orin and Oh [87] allows a sequence of joint torques to be determined off-line for a walking machine, but is too slow for real-time use. However, for a hand making small motions with the fingers, inertial terms can be ignored, considerably reducing the computation time. Power expenditure might actually be of little consequence in a robot gripper, but other terms might be added to the overall cost function to determine an optimum set of joint torques and stiffnesses for a given grasp geometry.

Mason [88] investigates the effects of friction on basic operations in which a robot grasps an object or pushes it into place. He points out that the role of friction in simple tasks performed by manipulators has not been adequately studied. The few investigators who have considered friction have been content to use the model developed by Coulomb in 1781. For tasks involving grippers and objects with hard, flat surfaces, the Coulomb model gives accurate results. Using it, Mason derives analytic solutions predicting, for example, the direction and the rate of rotation of an object pushed along a flat surface.

---

[2] In fact, a hand supporting a basketball is like a very large animal walking upon a very small planet.

For grippers with soft fingers (and particularly the human hand) the Coulomb model of friction may not accurately describe what we observe from experience:

> "When there is a possibility of the object slipping over the skin, a resistance, namely friction, intervenes which is proportional to the area of the surfaces in contact. ...Sweat glands, by moistening the skin, tend to increase friction and make the skin more adhesive." [89]

At light pressures, adhesion contributes greatly to the tangential force that a contact can sustain without slipping. The adhesion is not directly related to the normal force, but depends on surface chemistry, surface roughness, and the past history of normal forces. As an illustration, a compliant elastomer, once it has been pressed against the surface of an object, can often resist tangential loads even after the normal pressure is reduced to zero. The Coulomb coefficient of friction in this case would be infinite.

Fearing [90] and Wolter *et al* [91] discuss the stability of a two-dimensional object held in a two-fingered gripper. Fearing [90] considers elastic fingers with point-contact fingertips and adopts the Coulomb friction model. With these assumptions the problem becomes statically determinate. Using methods similar to those discussed by Mason [88], he determines conditions under which an object slipping against the fingers will come to rest in a stable position. He then considers whether a grasp will start to slip when disturbance forces and moments are applied to the object and looks at the problem of determining whether an object will remain stably grasped when manipulated between a moving and stationary finger. Wolter *et al* [91] discuss several algorithms for assessing the stability of prismatic shapes held in a parallel-jaw gripper with flat gripping faces. In particular, they consider resistance to slipping and the amount of work required to twist the object between the gripper jaws.

Tella *et al* [67] discuss methods for grasping parts heaped in a bin. A computer vision system determines tentative grasping orientations, and sensors on the grippers indicate whether a successful grasp has been established. The design and control of robotic hands has much in common with the design and control of teleoperators and hand prostheses [92-96]. Like robotic hands, these devices ideally require sensors for detecting slip, gripping forces and shear forces at the fingertips, and control algorithms to permit the hand to grip gently when desired.

CHAPTER 3
Robot Tasks in a Metal-Working Cell

An automated metal working cell is the focus of the present study, and the examples chosen to illustrate basic gripping and manipulation concerns are drawn exclusively from the metal working environment. However, similar concerns appear in many other robot applications.

Ideally, one would like a robot in a flexible metal working cell to do most of the tasks that people now do using their hands or hand tools. The tasks include transporting parts from one machine to another, loading them into fixtures and clamping them in place. To prepare for a new batch of parts, people may also assemble fixtures from basic components. With hand tools, people grind and polish parts, drill holes, rout panels and tighten screws and bolts on fixtures and workpieces. Most of the manual tasks within a metal working cell can be roughly grouped into three categories: materials-handling tasks, assembly tasks and surface finishing tasks.

3.1 Task Descriptions

3.1.1 Materials Handling

In materials handling tasks, robots transport parts from one machine to another, fetch them from conveyors or deposit them in racks as shown in Figures 3-1 and 3-2. The robot needs a large working envelope to achieve arbitrary positions and orientations within the cell and to reach around obstacles. Often the robot is a six-axis arm rooted to the floor, as in the case of the manipulator used for experiments in Chapter 4. Alternatively, the first axis may be a sliding track as in Figure 3-2. The robot does not have to maintain a precise trajectory as it travels, but since it moves over distances of several feet, the robot should be capable of speeds over 40 inches/second to be productive.

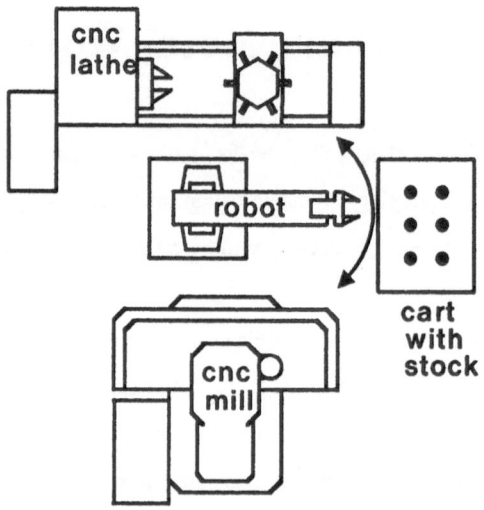

Figure 3-1:   Materials handling with large 6-axis manipulator (from [97])

The combination of high speeds and heavy workpieces places substantial power demands on the manipulator. A typical industrial robot for use in a metal working cell may have an overall capacity of 25 hp. With large actuators connected together in a serial kinematic chain, such robots are inherently gross-motion devices, suited for speed, flexibility and a large working envelope. However, their design is inherently unsuitable for such fine manipulation tasks as assembly and surface finishing.

3.1.2 Assembly

Many tasks within the cell involve elements of assembly. For example, when the cell switches from one part style to another, the fixtures on the machines have to be reconfigured. The robot assembles new fixtures from modular subcomponents and disassembles old fixtures. An assembly operation also takes place whenever a robot uses a power wrench or a screwdriver to fasten clamps or to screw fixturing components together. As the robot slips the wrench over the head of a bolt, or inserts the tip of the screwdriver into the head of a

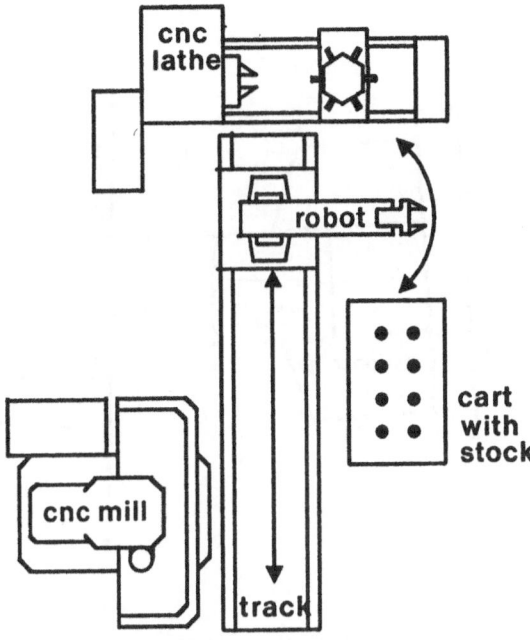

Figure 3-2: Materials handling with manipulator and track (from [97])

screw, it establishes a temporary assembly of tool and fastener. The robot performs still another assembly whenever it loads workpieces into clamps or fixtures on the machines in the cell. As the workpiece is slid into place, contact forces arise between it and the fixtures. This assembly operation is distinguished from the previous ones by the size and weight of the parts being assembled. Fasteners and fixturing pieces are generally small and light, but workpieces may be several inches across and weigh 100 lb or more. In all of the above assembly tasks, the robot must be compliant, responsive to changes in force and capable of fine movements. Since motions are small and velocities are low, there is no need for high power or a large working envelope. Dynamic terms are negligible, and a quasi-static model of the operation may be used. These are typical fine-manipulation characteristics. Finally, assembly tasks are not inherently time-dependent. As a result, velocities are unimportant; if the robot pauses during the assembly, no harm is done.

Figure 3-3 shows a robot attempting to insert a peg into a hole. The peg is a rigid body,

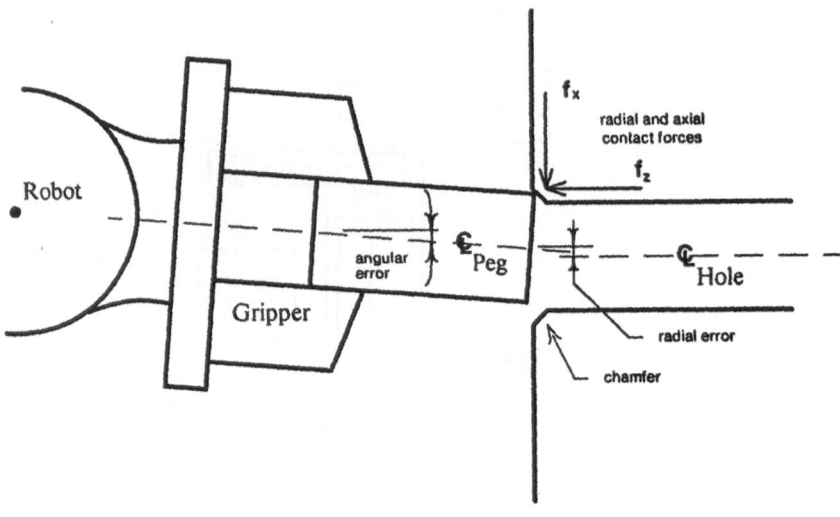

Figure 3-3: Beginning of insertion of peg in hole

but it is supported by a robot-wrist-gripper structure that has some compliance and determines its force/deflection behavior. Depending on the force/deflection behavior of the peg, the contact forces between the peg and the chamfer will guide the tip of the peg toward the center of the hole and cause the peg to tilt slightly. The effect of guiding the tip of the peg toward the center of the hole is helpful, but the tilting usually is not. If the peg initially is parallel to the axis of the hole, the tilt will produce an angular misalignment, as shown in the upper part of Figure 3-4. If the peg initially is misaligned, the tilt will either improve matters or make them worse, depending on which way the peg was leaning prior to contact. The safest thing is therefore to center the tip of the peg over the hole without tilting it. The contact forces can be made to do this if the force/deflection behavior of the peg is decoupled for forces applied at the tip. A point for which the force/deflection behavior of a structure becomes diagonal may be called a *center of compliance* [38]. At a center of compliance, the stiffness or compliance matrix of the structure becomes diagonal, and a force or torque in any given direction produces a deflection only in the same direction. Thus when inserting a peg into a hole, it is ideal if the center of compliance is at or near the tip of the peg.

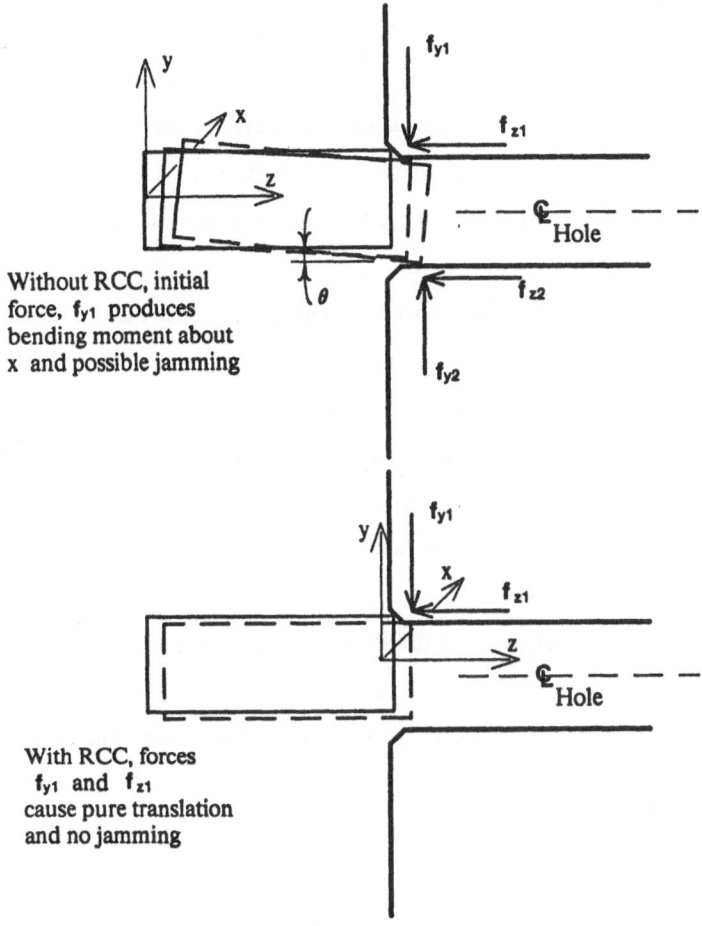

Without RCC, initial
force, f_{y1} produces
bending moment about
x and possible jamming

With RCC, forces
f_{y1} and f_{z1}
cause pure translation
and no jamming

Figure 3-4: Insertion with and without decoupled force-deflection behavior

After the peg is part way into the hole, contact forces between the peg and the hole produce a moment about the tip that will tend to further align it, provided that the coefficient of friction and the angular misalignment are not too large. It is still useful to decouple the force/deflection behavior of the peg so that moments about the tip of the peg will right it without side-effects. As the assembly progresses, the peg becomes constrained by the hole. The position and orientation of the peg in the x and y directions, perpendicular to the hole axis, are fixed. The robot is able to control only the motion of the peg along the z axis at the center of the hole, and the rotation of the peg about the z axis. Mason [18] has suggested that the constraint provided by the hole represents a *C-surface* with orthogonal force and velocity constraints: the robot cannot control the force along the hole and cannot control the position in the radial direction. Consequently, it is ideal if the robot is compliant in the radial directions (so that errors in the position of the robot do not produce large radial forces) but stiff in the direction along the hole so that the depth and velocity of insertion can be controlled precisely.

The basic techniques described above for inserting a rigid peg into a hole also apply to multiple peg assemblies and to the assembly of compliant parts. The diagonalization of the stiffness matrix has been achieved with passive compliant devices [38, 44], with active control [26, 98, 20, 39, 8, 22], or a combination of the two [38, 28, 29]. As discussed in Chapter 2, there are advantages to each method. However, in all cases, better performance is possible if the compliant accommodations are made with a small, light-weight manipulator or wrist instead of a heavy industrial robot.

To summarize, assembly operations are readily partitioned into two subtasks: one that moves the peg toward the hole and another that aligns the peg for assembly. The arm performs the first task and a compliant device performs the second. The two can function independently because:

- Assembly is quasi-static and independent of time. Therefore, the actions of the robot arm do not have to be synchronized with those of the compliant device.

- Assembly problems are represented by a *C-surface* in which compliant accommodations are orthogonal to the directions in which motion can be controlled. Therefore, the accommodations and the larger motions are physically uncoupled.

- For smooth assembly, the stiffness matrix of the peg should be diagonal near the tip of the peg. An RCC device produces the required matrix without any action on the part of the arm.

3.1.3 Grasping

Before a robot can transport or assemble parts, it has to pick them up. The robot may have to lift the part from a conveyor or to extract it from a set of clamps. The inability of industrial grippers to grasp a wide variety of parts has been recognized as one of the most serious limitations in today's manufacturing cells. The grasping problem can be broken into subtasks including: how to design better grippers, how to choose a suitable grasp for a given gripper and how to control a grasp once it is chosen. Most of today's grippers are designed on an *ad hoc* basis. Today's robot does not choose a suitable grip, but repeats the grips in its program. Control is reduced to "open gripper" and "close gripper" commands.

The metal working cell would be more self-sufficient if grippers were more versatile and if grasp planning were automated. Possible grips could be compared using a number of criteria including geometric constraints and mechanical properties. As an example of a geometric constraint, a part lying on a conveyor belt cannot be picked up by sliding the gripper beneath it, unless the gripper is very thin. Mechanical properties include the strength, stiffness and stability of the grip, and are explored in Chapter 5.

3.1.4 Surface Finishing and Shaping

Surface finishing and shaping tasks cover a wide range of operations from polishing surfaces with a wire brush to shaping them with a routing bit or an abrasive wheel. In these tasks, the robot usually holds a power tool and follows a contour on the workpiece, removing material as it goes.

In general, surface finishing and shaping tasks begin with an approximate description of the shape, position and orientation of the surface of the workpiece. The geometry of the workpiece may be provided by a CAD database and the position and orientation may be established by a computer vision system or by mounting the workpiece in fixtures whose positions are known with respect to the robot. From the description of the workpiece surface, an estimated path is computed off-line for the robot.

Once the estimated path has been computed, the task description varies depending on the nature of the finishing or shaping operation. The finishing and shaping tasks may be classified as soft or stiff, high speed or low speed and dependent or independent of time. Related tasks outside the metal-working cell include robotic arc-welding, caulking and painting, which also require a robot to use sensors in tracking a trajectory on the workpiece.

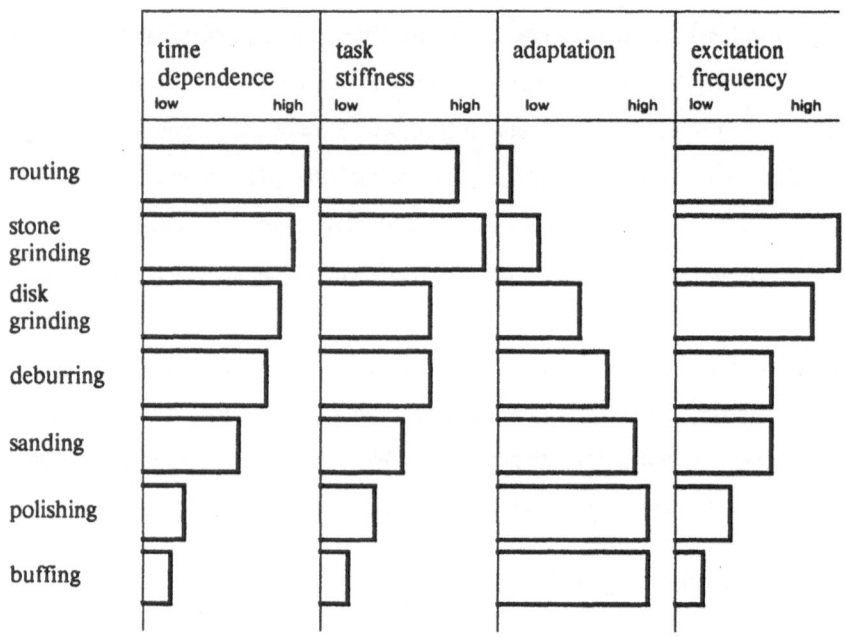

Figure 3-5: Properties of surface finishing tasks in a metal-working cell

- ***time dependence***

  Unlike the assembly operations described in Section 3.1.2, surface finishing or shaping operations may be time-dependent, since the tools will continue to remove material as long as they are in contact with the surface of the workpiece. The time dependence is least for polishing operations and greatest for heavy grinding. No harm is done if a robot slows down or pauses while it polishes a part with a wire brushing wheel. On the other hand, if the robot slows down or pauses while it grinds the edge of a casting, the grinding tool will gouge the surface. In between are tasks such as sanding; slowing down has little effect, but pausing produces circular marks on the object. The coupling between the robot velocity and the amount of material removed can sometimes be used to advantage since it permits the robot to alter the shape of the part as discussed in Section 3.1.4.2. Grinding and routing have much in common with robotic arc-welding because a welding robot must vary its speed to deposit more or less weld material as the weld seam widens and narrows.

- *adaptation*

Surface finishing operations require a mixture of positional accuracy and adaptability that depends on the amount of material being removed. In polishing and finishing, the robot follows the existing contour of an object, removing a fairly consistent amount of material as it goes. For example, if the robot is polishing a part with a buffing wheel, the appearance of the surface is changed but, since it removes very little material, undulations on the surface are preserved. It is important for the robot to be able to track unexpected bumps and undulations while producing a finish with a uniform appearance. As in the assembly tasks discussed in Section 3.1.2, the emphasis is on conforming to an existing geometry. A number of techniques are available to permit the robot to track a surface and are discussed below in Section 3.1.4.1. In sanding and grinding operations the same approach may apply, but the amount of material removed will be greater for bumps on the surface and less for valleys. The result is that the contour of the workpiece becomes smoother. In routing and abrasive cutting, the goal is to establish a smooth contour independent of the bumps on the surface of the object.

- *stiffness*

Buffing and wire brushing tasks are soft because the tool is deformable. As a result, minor positional errors between the robot trajectory and the surface of the workpiece do not produce large changes in the force applied to the surface or in the torque required from the tool motor. Grinding and routing tasks are much stiffer; the tools are essentially rigid and position errors between the robot and the workpiece produce large variations in both the force against the surface and the grinding torque. In between are tasks in which the robot sands or grinds the surface using a fibre reinforced disk. Outside the metal working cell, the ultimate example of a "soft" task is painting. In painting the robot may use an optical or ultrasonic sensor to maintain a constant distance from the surface, but since there is no direct contact between it and the workpiece the task stiffness is zero.

- *characteristic frequency*

For the most part, wire brushes and polishing or buffing tools are driven at low speeds (under 2000 rpm) and, as a result, the task induced forces and torques transmitted to the robot have a characteristic frequency of 200 Hz or less. Grinding, however, may involve speeds of over 15,000 rpm and the forces transmitted to the robot may have frequencies of over 1200 Hz. Sanding and routing involve intermediate speeds, with characteristic frequencies of several hundred Hertz. In practice, the periodic forces produced by these tasks do not usually affect the dynamics of a large industrial robot, which frequently has a bandwidth of just a few Hertz. Fortunately, the polishing and buffing tasks, which have the lowest frequencies, are also the softest tasks with the smallest forces. An exception to the above rule has been found in robotic de-finning of castings [52]. In the de-finning of castings, a pneumatic chipping hammer is used to remove the flash around the edges of cast iron workpieces. Forces generated by the hammer have peak values of up to 3000 lbf and a frequency of 42 Hz.

3.1.4.1 Contour following

In most of the surface finishing and shaping tasks described above, it is important for the robot to be able to follow the contour of a workpiece. Even for tasks in which the robot substantially defines the final shape of the workpiece, it is often useful for the robot to first track the contour, with the tool turned off, to establish the exact location and orientation of the workpiece and to locate bumps on the surface that will have to be removed.

Like the assembly tasks discussed above, contour following requires the robot to conform to a given geometry. More explicitly, the robot should be compliant in the direction normal to the surface of the object, so that unexpected variations in the surface do not produce large changes in the force that the robot applies against the surface. Variations in the force against the surface are undesirable for several reasons: they may squash a wire brush or excessively bend a flexible sanding disk, they may overload the motor of the finishing tool and they generally produce an uneven surface appearance. In the directions parallel to surface, the robot should not be compliant; it needs to maintain a desired trajectory and should therefore be position-controlled. The result is that the workpiece can be considered a *C-surface* [18] which constrains forces in the tangent plane and constrains the velocity perpendicular to the tangent plane. As in assembly, the orthogonal force and motion constraints permit the use of independent systems for positioning and compliance.

If it is known that the actual trajectory traced by the robot will deviate only slightly from the estimated path, compliance techniques are sufficient for the robot to cope with errors. However, when the off-line estimate of the robot path is less certain, the robot must use sensors to track the contour, continuously updating the estimate for the remainder of its path. A typical tracking task is illustrated in Figure 3-6 in which a robot pushes a round tool, such as a polishing wheel, along an irregular surface. For tracking the contour, it is useful to distinguish between long-term and short-term errors:

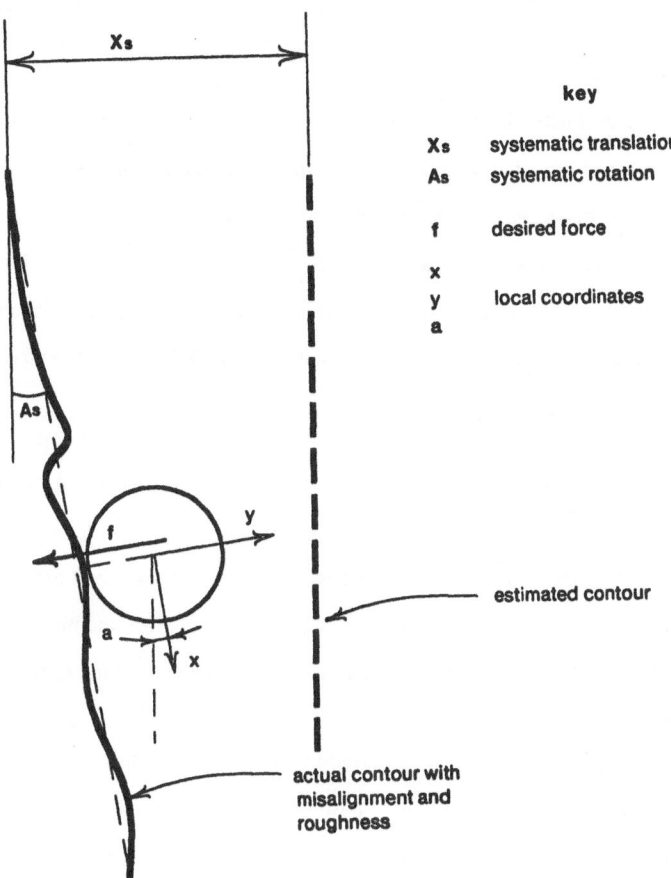

Figure 3-6: Tracking a contour with misalignment and roughness (from [29])

- Long-term errors show a statistically significant trend over a period of time. They may result from misalignment of the workpiece, from robot miscalibration or from errors in the CAD description of the part. Figure 3-6 shows a long-term error due to misalignment of the workpiece. Long-term errors must be accounted for so that the robot can update its predicted path.

- Short-term disturbances are also present in the task. As the robot follows the contour it is subjected to fluctuating forces and torques resulting from local variations in the coefficient of friction between the tool and the workpiece and from the actions of the tool as it encounters soft or hard material. In addition to force/torque disturbances there will be position disturbances due to surface roughness and due to the limited resolution of the robot. Figure 3-6 shows the effects of surface roughness. These essentially random fluctuations are superposed on the long-term errors described above.

The random fluctuations are unavoidable; they represent the resolution of the robot and the magnitude of disturbances inherent in the task. The robot must be sufficiently compliant to absorb these disturbances without producing excessive contact forces between the tool and the surface. The task of the contour-following algorithm is to extract the long-term trends from a mixture of data and to modify the robot path accordingly. In Section 4.3, methods are investigated for tracking a surface using sensory data from an instrumented robot wrist.

3.1.4.2 Contour modification

For heavy grinding, abrasive cutting and routing, it is desirable not just to smooth the surface, but to reduce the undulations of the surface, producing a smoother contour. For example, forged parts often have wavy edges as a result of excess metal being extruded from between the forging dies. The waves must be removed, leaving a smooth edge. In aerospace applications, large panels often must be routed to produce a finished edge for mating with adjoining panels on an airframe.

For tasks such as these, where the requirement is to modify the geometry of the part, and not merely to alter the surface finish, positional accuracy becomes important. Positional accuracy is not the strong point of large industrial manipulators. As open kinematic chains they are designed for flexibility and versatility. Precision grinding and routing are more commonly done using CNC machine tools. However, it is convenient to use robots when the parts are large or rough (and therefore difficult to mount in fixtures on a machine tool), where it is desirable to permit some uncertainty in how the part is held and where tolerances do not need to be as close as those obtained on machine tools.

When a machine tool routs a panel or cuts it with an end mill, it moves in a fixed trajectory. Variations in the depth of cut, the feedrate or material properties may produce large variations in the cutting or grinding force, but because the machine tool is a very stiff, position-controlled device, the variations do not cause it to deviate from its planned path. The resulting part is accurately shaped, but care must be taken to prevent excessive forces from breaking or prematurely wearing the cutting or grinding tool, from producing a poor finish or from shifting the part in its fixturing. A machinist working on a manual machine tool continually monitors the cutting process and adjusts the feedrate to prevent excessive forces from developing, but on CNC mills, a feedrate is chosen that will be safe for all cutting conditions expected during the task. In an effort to emulate some of the actions of a skilled machinist, adaptive control schemes have been proposed for milling and grinding [99, 100, 101, 102]. The feedrate may be continuously varied to keep the cutting force within specified limits. This approach extends the life of the tool, reduces the likelihood of chatter and may increase the dimensional accuracy that can be obtained, particularly when flexible cutting tools are used [99].

A similar approach is investigated with a large industrial robot in Section 4.3. The velocity at which the robot moves over the surface of the workpiece is analogous to the feedrate of a CNC milling machine. If the robot maintains a constant force normal to the object surface then, since the grinding or routing task is time dependent, varying the velocity produces variations in the depth cut. Thus, as the robot approaches a large bump on the surface of the object, there are two options:

- The robot can continue at the same velocity over the surface and with a constant force normal to the surface, climbing over the bump and removing a uniform layer of material.

- The robot can slow down so that the grinder cuts deeper into the bump, partially or completely grinding it off. As the grinder passes the bump, the robot resumes its original velocity.

In the adaptive control of machine tools, and in the compliant surface finishing tasks described in Section 3.1.4.1, it is desirable to control the force against the surface to avoid overloading the tool motor, to prevent premature tool wear, and to avoid tool chatter. In surface-shaping tasks, maintaining a consistent force also improves the positional accuracy of the robot. Industrial robots are flexible kinematic chains that deflect considerably when forces are applied to the hand, resulting in a deterioration in positional accuracy. Methods for improving the dynamic stiffness and accuracy of robots are reviewed in Chapter 2, but in general, positional accuracy is more readily achieved under consistent loads than under loads that vary widely and in an unpredictable manner.

3.2 Discussion: Coupled Fine and Gross Motions

When a robot picks up a peg and pushes it into a hole, or follows a rough surface, orthogonal directions can be found in which the robot should be either compliant or stiff. Where the assembled parts or the surface constrain the robot, it is limited to small, compliant motions. Where the robot is unconstrained, it makes larger motions and may be stiff. Such tasks are also independent of time; if the robot pauses, no harm is done (apart from a reduction in productivity.) In other words, the fine and large motion elements of assembly and contour following tasks can be broken into *orthogonal, unsynchronized* and *uncoupled* elements. As a result, independent systems, consisting of a robot arm and an RCC device can be used to accomplish the large motion and fine motion subtasks respectively.

In surface finishing and shaping there are also fine and large motion elements, but they are coupled because material is removed from the workpiece. This adds an element of time-dependence, and the difficulty of the tasks is proportional to the rate of material removal. In polishing and sanding tasks, the robot removes a fairly uniform amount of material from the workpiece and can proceed at a constant velocity along the surface while maintaining a constant force against it. A difficulty arises, however, when little is known about the workpiece. In this case the robot must track the surface, continuously reassessing the surface orientation and determining directions for compliant and stiff behavior.

In coarse grinding and routing tasks, the directions for compliant and stiff robot behavior are less distinct. The robot modifies the shape of the workpiece as it travels. Under constant force conditions, the velocity of the robot is inversely related to the amount of material removed. Under constant velocity conditions, the forces increase dramatically as the depth of cut increases. A constant-velocity approach is traditionally used when CNC machine tools grind and rout parts. The tools are sufficiently stiff that variations in force do not produce large positional errors. Industrial robots in dedicated grinding operations are also generally slower, stiffer and more accurate than the manipulators used in manufacturing cells. However, constant force methods, or perhaps a combination of force and velocity variations, may permit flexible, general-purpose robots to attempt coarse grinding.

In the following chapter, a wrist and robot arm are investigated for assembly, surface finishing and surface shaping tasks. The wrist achieves the fine accommodations required in assembly and surface finishing. For tracking, the wrist must communicate with the robot, providing filtered sensory information about the local orientation of the surface. For grinding, a higher level of communication is required since the robot velocity must be adjusted using sensory information from the wrist, and the optimum wrist compliance may vary as a function of robot velocity.

CHAPTER 4
A Wrist
for Fine-Motion Tasks

When humans perform manufacturing tasks requiring both large and fine motions, for example when using a large wrench to tighten a bolt, they use arm muscles for achieving the required speed and power, but use wrist muscles for the fine accommodations needed to slip the wrench over the head of the bolt. During the fine motions, the arm serves as a stable support for the wrist and hand. Lighter tasks, such as writing with a pencil or assembling small parts, may use finger manipulation. However, in a metal working cell, tools and parts are often too heavy for humans to manipulate with their fingers. Instead, people grasp the parts with a passive grip in which the fingers conform to the tool or object, and they make motions with the wrist and arm. Likewise, an active wrist may be combined with a conventional, passive gripper in many manufacturing tasks.

4.1 Wrist Description

The wrist developed for present work is a compliant unit with sensors and spherical hydraulic actuators that behave as variable-stiffness springs. Like the RCC devices reviewed in Chapter 2, the wrist has a stiffness matrix that becomes diagonal at a point some distance in front of the wrist. However, since the spherical "springs" are adjustable, the location of the center of compliance can be matched to the size of a given tool or workpiece. With a tunable center of compliance, the wrist achieves the fine accommodations needed in assembly and contour following tasks. For tasks such as grinding, which require force/torque feedback, the sensors measure deflections in the wrist. A dedicated microprocessor takes care of the details of monitoring the sensors and adjusting the spheres. The microprocessor forms part of a distributed control system in which the other partners are a task controller and the robot arm controller.

The elements of the wrist are reviewed below. A more detailed discussion of the mechanical design can be found in [28, 103]. A description of the control system and a discussion of the calibration of the wrist are given in [29].

Mechanical design

A design drawing of the wrist is shown in Figure 4-1. Like other RCC wrists in [42, 43] and the assembly device in [44], the wrist is an elastic linkage. Figure 4-2 shows a idealized planar schematic of the wrist. The base of the wrist is bolted directly to the end of a manipulator and the upper plate of the wrist is bolted to the gripper so that the gripper "floats" relative to the arm. Between the base and the upper plate are several elastomeric spheres which function as springs, and a number of cables which hold the base and the upper plate together in tension. Axial spheres give the wrist stiffness in compression and bending while radial spheres give the wrist stiffness against lateral (radial) loads.

In Figure 4-1, the axial spheres and cables are inclined toward the central z axis of the wrist by a slight angle, θ. The inclination of the cables and spheres allows the center of compliance to be projected a distance l_2 from the upper platform. The axial spheres are made of hollow, reinforced rubber spheres. Their stiffness, k_{yb}, is changed by adjusting the pressure of a fluid within them. This changes the stiffness of the wrist and alters the projected distance, l_2, of the center of compliance. The cables are made of Kevlar[3] yarns. The yarns are very flexible, but have an elastic modulus and tensile strength comparable to steel. As a result, the cables function essentially as pinned links in tension, but have no effect in compression. The cables permit a higher ratio of bending stiffness to compressive stiffness than is achieved in most other *RCC* devices. As a result, the wrist is better able to withstand cantilevered loads than other *RCC* devices. The price paid for this feature is that the wrist can be compressed but not stretched and consequently has only 1/2 degree of freedom along the z axis. However, compliance in extension is required far less often that compliance in compression; the robot usually requires compliance as it pushes into objects and not as it pulls away from them.

For the two-dimensional linkage shown in Fig 4-1, no rotation occurs if the lateral force, f_x, is applied at a distance

---

[3]Kevlar is a trademark of Dupont Inc.

34

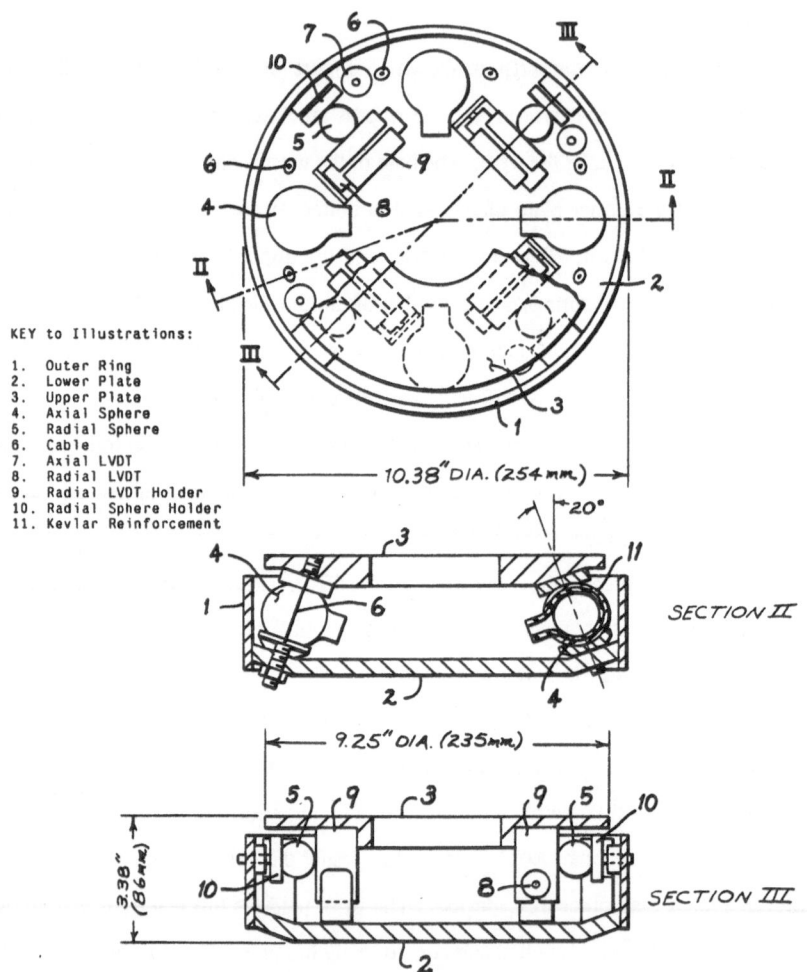

KEY to Illustrations:

1. Outer Ring
2. Lower Plate
3. Upper Plate
4. Axial Sphere
5. Radial Sphere
6. Cable
7. Axial LVDT
8. Radial LVDT
9. Radial LVDT Holder
10. Radial Sphere Holder
11. Kevlar Reinforcement

Figure 4-1: Design of Compliant Wrist (adapted from [103])

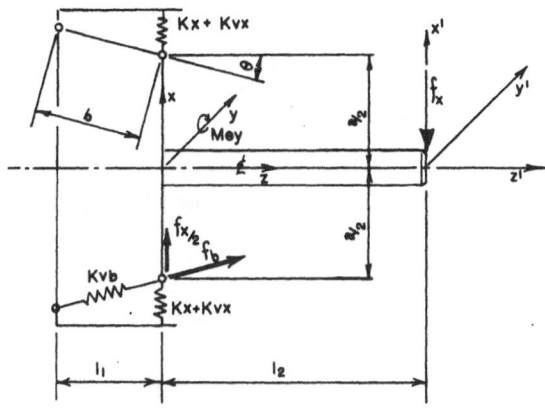

Figure 4-2: RCC Linkage (from [28])

$$l_2 = \frac{a\,k_{yb}\,\tan\theta}{k_x + 2k_{yb}\,\tan^2\theta} \tag{4.1}$$

from the upper platform. This is the projected distance of the center of compliance. For the actual device, with four springs and cables instead of two, the three-dimensional projection of the compliant center becomes

$$l_2 = \frac{3/2\,a\,k_{yb}\,\tan\theta}{k_x + 3\,k_{yb}\,\tan^2\theta} \tag{4.2}$$

where k_{yb} is the stiffness of each axial sphere, k_x is the stiffness of each radial sphere, θ is the angle of inclination of the axial spheres and cables and a is the diameter of the wrist.

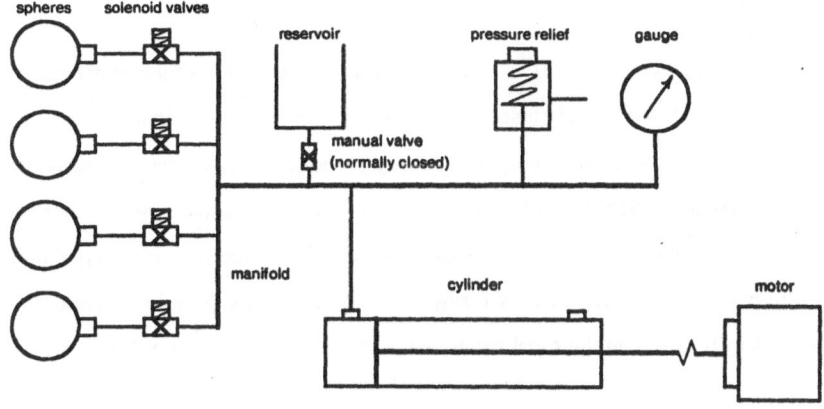

Figure 4-3: Hydraulic Schematic of Wrist (from [29])

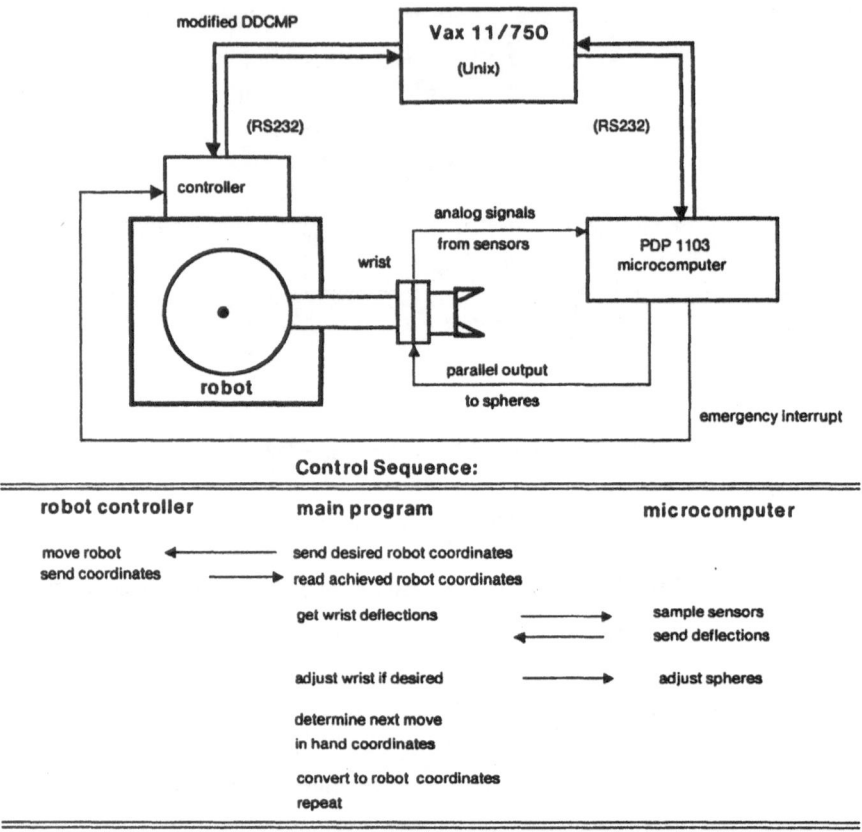

Figure 4-4: Communications for robot sensory control
(modified from [29])

Instrumentation

The wrist contains eight LVDT[4] sensors that measure the motion of the upper platform with respect to the lower. As discussed in [29], the accuracy of the wrist is limited primarily by small-angle approximations and by the linearity of the displacement/voltage signal from the sensors. When the wrist is calibrated, the *rms* errors are approximately 0.0003 inches for small translations and 0.0003 radians for small rotations. When the wrist is mounted on the robot arm in experiments the accuracy is slightly lower because the position and orientation of the wrist with respect to the robot cannot be perfectly established. In practice, the wrist

---

[4]Linear Variable Differential Transformer

remains reliably within 0.001 inch for all but large deflections of 0.11 inches or more. For calibration, the wrist is placed on a CNC machine tool, which moves the upper platform while the sensor readings are recorded and used to determine the least-squares pseudo-inverse of the calibration matrix.

Hydraulic system

The hydraulic system for controlling the wrist is illustrated in Figure 4-3. The wrist is not a servo-controlled device, but the fluid volume in each sphere can be adjusted once or twice a second. The adjustment of the spheres is essentially multiplexed, using a single motor and piston for adding or removing fluid. The result is a system that cannot be dynamically controlled in the way that the wrists in [39, 40, 41] can, but which is considerably simpler and more robust and which can be "tuned" to match the requirements of a task. The wrist is adequate for assembly, surface finishing and surface shaping experiments in which the requirements of the wrist are that it exhibit specified compliant characteristics and provide sensory data for controlling the robot.

4.2 Control Architecture

The control scheme for the arm and wrist is shown in Figure 4-4 and the individual elements are discussed below. The control system is hierarchical, with the task controller directing the arm and the wrist controllers.

Task controller

The task controller currently runs on a VAX 11/750 (the software development computer for the cell) but could just as easily run on a microcomputer with disc storage and a floating point processor. The controller runs programs for predicting the trajectory and for determining the current state of the robot and wrist. It also generates commands for the robot and wrist controllers. At each step the task controller may send the robot a new velocity, new coordinates and instructions to open or close the gripper or turn on a power tool. The task controller may ask the wrist to measure deflections and to adjust the pressure of the spheres. The spheres may all be given the same pressure (as is usually the case when the central axis of the wrist is aligned vertically) or different pressures to allow for static loads. The ability to accommodate static bending loads is especially useful when the central axis of the wrist and gripper are horizontal. As the robot begins a new task, the task controller may use data from the wrist to correct the initial pressure settings. For example, if the sensors show that the wrist is tilting or bending when contact forces are ostensibly acting at the center of compliance, the task controller directs the wrist to become more or less stiff until the tilting disappears. It is also possible to direct the wrist to rock slowly from side to side by alternating the pressures of spheres on opposite sides of the wrist.

Wrist controller

The wrist controller reads and smooths the signals for the eight sensors and converts them to deflections in inches. If any of the sensor readings fall outside a specified range, the wrist sends an interrupt directly to the robot controller to abort the task. The range is currently specified at compile time and corresponds to the largest expected wrist deflections for the task. In normal operation, the wrist controller sends the deflections to the task controller as a six-element vector in cartesian hand coordinates. The wrist controller also interprets command strings from the task controller telling it how to adjust the wrist.

A more powerful wrist controller, with better floating point capability and some non-volatile memory, could make low-level decisions (currently made by the task controller) to adjust the spheres. The wrist would be given a set of estimated stiffness properties required for the task and a pattern of sensor readings that it should expect. The wrist controller would then readjust the spheres in response to sensor readings. The decisions made by the wrist controller could always be over-ruled by the task controller.

Arm controller

Exchanging coordinates between the task and robot controllers is presently the most time consuming part of the control sequence shown in Figure 4-4. At present, the arm controller only accepts commands over a serial line. The maximum velocity between points can be specified, but continuity of velocity at the points is not ensured. As discussed in Section 4.3.2, this causes difficulty with time-dependent tasks such as grinding.

4.3 Experiments

Figure 4-5 shows the compliant wrist inserting a round peg into a hole. As discussed in Section 3, the wrist should be able to perform assembly tasks using the RCC effect without sensory feedback and without communicating with the robot controller. This was verified with the wrist for pegs between 1 and 4 inches in length. Similarly, following a rough contour can be achieved using compliance and no feedback — provided that the deviation between the expected and actual contour never exceeds the compliant range of the wrist and robot. However, when the uncertainties in the contour are large, the robot must use sensory feedback to track the contour, continually updating its estimate of the orientation of the surface. When the robot is used in surface shaping tasks, as discussed in Section 3.1.4.2, the robot again requires sensory feedback in addition to compliance. Surface finishing and shaping were therefore selected for further investigation.

Figure 4-5: The instrumented wrist inserting a peg into a hole

4.3.1 Contour Following

As discussed in Section 3.1.4, the robot begins a tracking or surface finishing task with some estimate of the path it will follow. The amount of information in the estimate depends on how well defined the shape of the part is and on how accurately the position and orientation of the part are known. If the part is described in a CAD database, the trajectory may be given as a series of straight lines, circular arcs and splines, subject to errors in the position of the part and manufacturing tolerances. In this case, the problem becomes one of fitting the coordinates measured by the robot and wrist to the CAD description. For simple shapes, it may only be necessary to touch the part in few places. Once the position and orientation have been established, the robot may be able to follow the contour using only compliance techniques. However, when the part geometry is ill-defined, the robot may begin with no more than a safe starting position, some information about the expected complexity of the trajectory (for example, the amount of curvature and roughness to be expected) and the general direction in which to start moving.

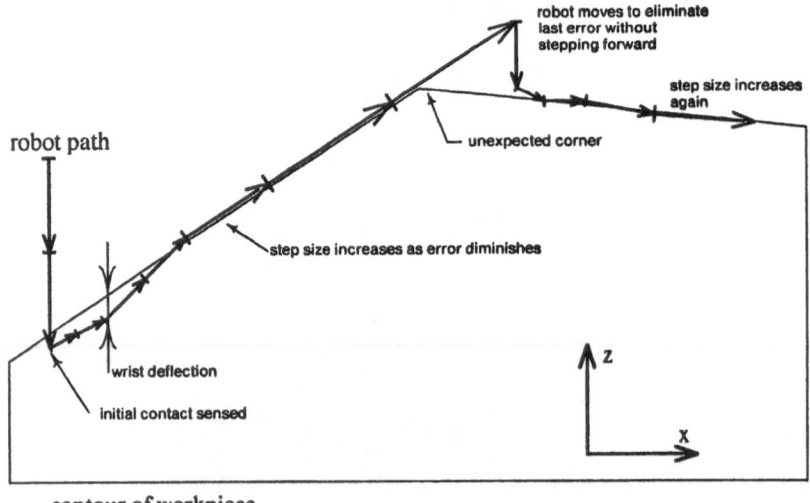

Figure 4-6: Variable-step tracking of a contour with an unexpected corner

Before going into the details of the filtering and prediction methods and experimental results, the main features of the tracking task will be discussed.

- **Procedure**
  The robot proceeds toward the surface until the wrist senses that it has made contact. The wrist signals the contact to the supervisory program, which starts the robot moving in the estimated direction of travel. At each interval, the sensor readings from the wrist are gathered, analyzed and used to specify the next robot move.

- **Time independence and arc length**
  A discussed in Section 3.1.4.1, a pure tracking task is relatively time-independent. The robot may be polishing the surface with a wire brush or it may be learning the shape of an object as the first step in a shaping operation, such as routing. In either case, there is little need to control the velocity of the robot. Consequently, it is convenient to use the arc length, s, along the trajectory as the independent variable instead of time.

- **Range of acceptable wrist deflections**
  The robot path is adjusted to keep the wrist deflections within a "window" of values, corresponding to a range of acceptable contact forces. If the robot is performing a surface finishing operation, such as polishing with a wire brush, the "window" ensures that the brush is held against the surface with a suitable polishing force. If the robot is measuring the contour for a later grinding operation, the wrist deflections are made to match the average deflections that will occur during grinding. This helps to keep the grinding forces from producing an error in the robot trajectory. With this method, the positions recorded by the robot may be inaccurate with resect to the external world, but because they are duplicated during grinding, they do not affect the shape of the workpiece.

- **Variable step size for efficient tracking**
  The process of transmitting wrist deflections to the supervisory program, sending motion commands to the robot and recording the robot position is relatively slow. A higher bandwidth communications scheme would reduce the turn-around time, but in general, the price that we pay for a distributed system, with separate arm, wrist and task controllers, is that communication among them limits the frequency with which the robot can respond to sensory signals at the wrist. With such a system, it is therefore important to have efficient tracking algorithms that permit the robot to move as far between sampling intervals as possible. Thus, the control algorithm should predict the direction of the surface as far ahead as conditions permit. In practice, the maximum "safe" prediction depends on how broad the window of acceptable wrist deflections is, on how

accurately the control algorithm can predict the surface ahead and how accurately the robot can move in response to commands. The window of acceptable deflections is given at the start of the task. The robot accuracy and the prediction accuracy are estimated initially, using a conservative guess. Once the tracking process has started, the recent history of errors (deviations between the ideal and measured wrist deflections) determines how far we are willing to move the robot in the next step. The result is that if the measured errors grow small, the robot moves in large steps and completes the tracking task rapidly. If the errors grow large, the robot starts to take smaller steps until the tracking algorithm has identified the new curve direction. If the measured error is outside the window of acceptable wrist deflections, the robot does not step forward, but moves directly into or away from the surface to eliminate the error. The effect is demonstrated in Figure 4-6. As the robot begins tracking, it takes small steps until the orientation of the contour is established. As the measured errors become small, the robot takes larger steps. When the robot encounters an unexpected corner, the wrist measures a large error and the robot moves directly toward the contour to eliminate the error. The robot then continues with steps that grow larger as the new direction of the contour is identified.

- **Sources of error**

As discussed in Section 3 there will be errors of two kinds: short term and long term. The robot is unable to change course rapidly enough to eliminate the short term errors but should respond to the long term errors. The compliance in the wrist absorbs the short term errors and prevents them from producing large changes in force. The task of the robot is to keep the wrist near the center of its compliant envelope. The wrist sensors are unable to distinguish between the different sources of error including surface roughness, robot inaccuracy and workpiece misalignment. Extracting long-term trends from sensor data is a filtering problem in which random fluctuations are thought of as "noise" corrupting a systematic signal. Whitney and Junkel [104] describe similar problems involving noisy sensor data and discuss the advantages of using a Kalman filter to extract the systematic trends.

4.3.1.1 State estimation

Nomenclature for estimator example

x Horizontal dimension

k Step counter

y Curve height above some reference line

v dy/dx

r	Robot height above some reference line
s	The state variable estimate vector in (c, v, r) after measurement
$\bar{s}$	The state variable estimate vector in (c, v, r) before measurement
w	Vector of process noise for (c, v, r)
Φ	The state transition matrix for s
Γu	The control law used for moving the robot in response to deflections (for example, proportional feedback)
Q	The process noise covariance matrix
R	The sensor noise covariance matrix
z	Vector of sensor readings from robot and wrist
n	Vector of sensor noise
H	Matrix relating sensor readings, z, to state variables, s
V	The Kalman filter signal vector matrix after measurement
$\bar{V}$	The Kalman filter signal vector matrix before measurement
G	The Kalman filter gain matrix

A Kalman filter and a finite-memory linear predictor have been used in tracking experiments with the compliant wrist. The concepts can be illustrated with a one-dimensional example. Without any prior knowledge about the contour, the prediction for the next point might be estimated according to the formula $y_{(x+\delta x)} = y_{(x)} + v\,\delta x$. The robot returns a value, $r_{(x)}$, at each step and the wrist returns a deflection, $d_{(x)}$. The nominal position of the curve is given by $y_{(x)} = r_{(x)} - d_{(x)}$. However, the values returned by the robot and the wrist are subject to random errors or noise. The noise in r is given by e_r and is due largely to the limited resolution of positions obtained from the robot joints. The much smaller noise in the wrist deflection is ε_s.

If the k is the counter for each time step, the state equations for the tracking system become

$$s_{k+1} = \Phi s_k + \Gamma u_k + w \qquad (4.3)$$

or

$$\begin{bmatrix} y_{k+1} \\ v_{k+1} \\ y_{k+1} \end{bmatrix} = \begin{bmatrix} 1 & \delta x & 0 \\ 0 & 1 & 0 \\ 0 & 0 & 1 \end{bmatrix} \begin{bmatrix} y_k \\ v_k \\ r_k \end{bmatrix} + \begin{bmatrix} 0 \\ 0 \\ 1 \end{bmatrix} u_{(y.v.r)} + \begin{bmatrix} w_y \\ w_v \\ w_r \end{bmatrix}$$

$$z_k = H s_k + e \qquad (4.4)$$

or

$$\begin{bmatrix} z_1 \\ z_2 \end{bmatrix} = \begin{bmatrix} 1 & 0 & -1 \\ 0 & 0 & 1 \end{bmatrix} \begin{bmatrix} y_k \\ v_k \\ r_k \end{bmatrix} + \begin{bmatrix} e_s \\ e_r \end{bmatrix}$$

w contains process noise terms for c, v and r. The noise in r stems from the limited resolution of the robot (the smallest distance the robot can move is about 0.25 inches) and from servoing errors. The noises in c and v result from surface roughness on the contour and force/torque fluctuations as the tool encounters harder or softer material and higher or lower coefficients of friction.

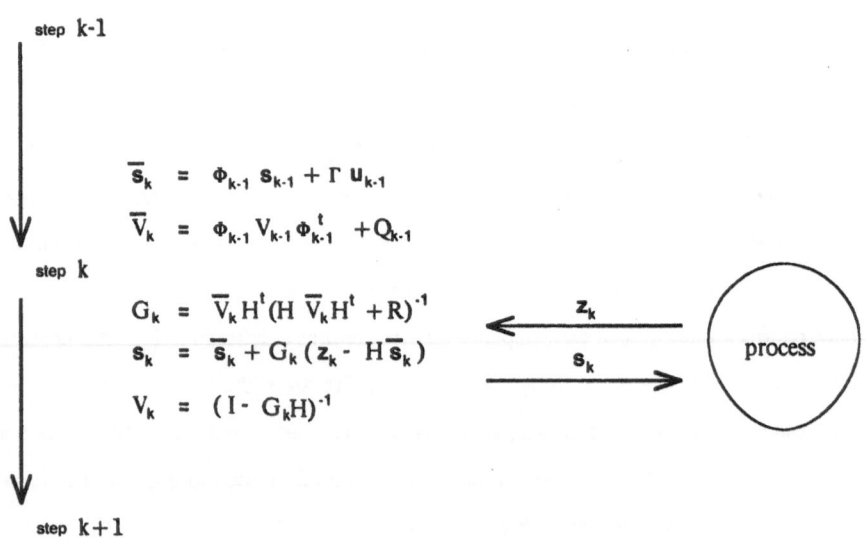

Figure 4-7: Implementation of Kalman filter estimator

The filtering program used in the tracking experiments is based on the Kalman filter equations presented in [105, 106, 107, 108]. The Kalman filter may be used either as a predictor, predicting what the height of the curve will be at the next step, or as a 0-step filter, providing an optimal estimate of the *current* curve location. In experiments, it has been found most efficient to use the Kalman filter as an estimator or observer and to use a separate, and simpler, predictor in the control algorithm as discussed below. For a time-invariant Kalman filter, it is assumed that the state transition matrix, Φ, and the noise properties are constant for the duration of the task. This necessitates a constant step length, δx, which is not ideal. Consequently a time-varying implementation is used in which Φ is updated with each step. Table 4-7 shows the timing of the equations used in the 0-step estimator program.

In the above equations, Q_k is the covariance matrix of process noise, w, in (c,v,r). In many cases Q can be assumed constant over the duration of a task. If the noise sources can also be assumed uncorrelated and zero-mean, Q becomes

$$Q = \begin{matrix} \sigma^2_y & 0 & 0 \\ 0 & \sigma^2_v & 0 \\ 0 & 0 & \sigma^2_r \end{matrix}$$

where σ^2_y, σ^2_v and σ^2_r are the expected variances in y, v and r.

The sensor noise is represented by the covariance matrix R, which is generally constant and may be expressed as

$$R = \begin{matrix} \sigma^2_{es} & 0 \\ 0 & \sigma^2_{er} \end{matrix}$$

where σ^2_{es} is the variance in wrist sensor readings and σ^2_{er} is the variance in robot sensor readings. In practice, the variance in the wrist sensor readings is negligible compared to the other noise terms since the wrist is more than an order of magnitude more accurate than the robot.

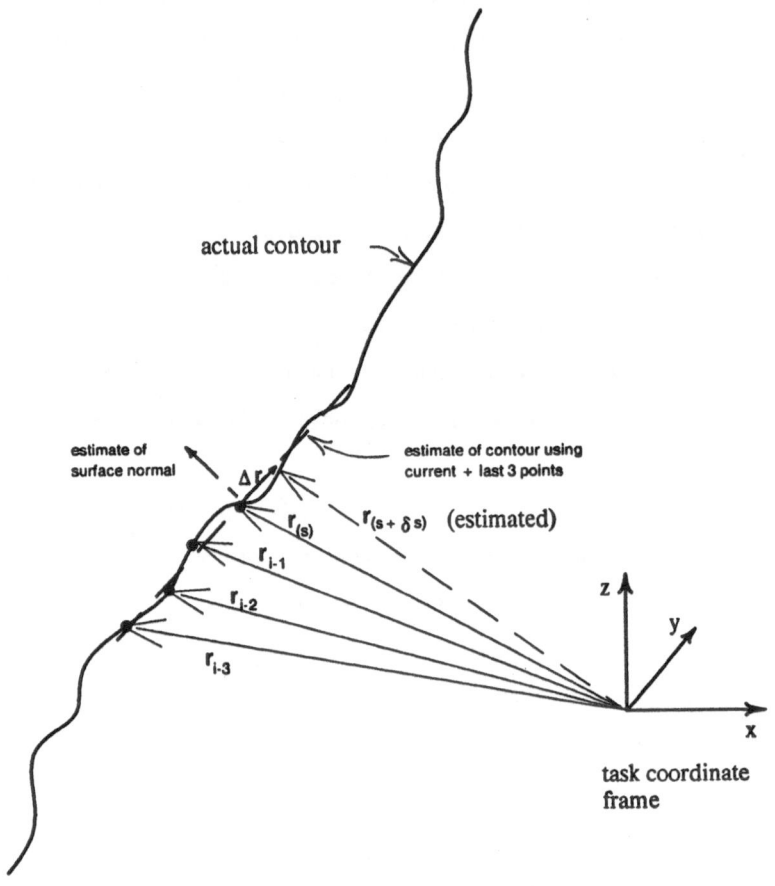

Figure 4-8: Least-Squares Prediction of Parametric Trajectory

4.3.1.2 Control law

Once filtered, the curve estimate, $c_{(s)}$, is used for controlling the robot and for updating the robot trajectory. The control algorithm could use pure proportional control but better results are obtained by using a predictor to estimate where the curve will be at the next time step. The problem with pure proportional control is shown in Figure 4-9. The robot never "learns" about the misalignment of the workpiece and continues to move straight ahead whenever the current error is small. The result is a staircase pattern of horizontal motions and vertical corrections. As discussed earlier, for efficiency, it is desirable for the robot to move as far between measuring intervals as possible, and the control algorithm determines how far it is willing to predict the curve ahead based on how large the current tracking errors are. With the large errors encountered using pure proportional control, the robot must take small steps forward and, in fact, it requires 49 steps to complete a 9 inch section of contour. Of the 49 steps, only 20 result in deflections that are within the specified window of 0.075 ± 0.015 inches. The problem is exacerbated by the resolution of the robot. The robot cannot reliably move less than 0.025 inches in any direction. Thus, if the robot moves a very small amount in the horizontal direction, the motion in the vertical direction is only accurate to the nearest 0.025 inch.

With a two-point predictor, the robot estimates the instantaneous orientation of the curve using the position of the current step and that of the previous step. Unfortunately, as Figure 4-10 shows, a two-point predictor is vulnerable to noise from errors in the robot coordinates. For a contour that is nearly straight, much better results are obtained using a recursive least-squares estimate based on the last several points. Figure 4-11 shows the results using the last 10 points. Because the measured tracking errors are small, the robot moves ahead in larger steps and completes the contour in only 27 steps. Of the 27 steps, only two result in deflections outside the window. The two unacceptable deflections occur at the start of the contour, when only a few data points are available for making a prediction. Thus, with a suitable predictor, the tracking accuracy approaches the resolution of the robot.

Since the predictor makes an estimate based upon the last few points, the effect is similar to adding integral control to the proportional control. In either case, the current command for the robot is a function of previous sensory information in addition to the current reading. An advantage to the predictor is that it directly reflects the dynamics of the curve. Thus, if the curve is locally approximated by a straight line, a linear predictor can be used; if the curve is locally approximated by a section of constant curvature a quadratic can be used, and so on. In the experiments with the compliant wrist, the surface curvature was generally small over several steps and therefore a linear n-point predictor was used.

Nomenclature for predictor

s_n $\qquad$ = arc length at n^{th} step

$\bar{s}_n$ = average arc length for n steps

r_n = (x,y,z) position at n^{th} step

$\bar{r}_n$ = average position for n steps

f_n = vector of s and r terms at n^{th} step

$\bar{f}_n$ = average of f for n steps

The predictor provides a recursive least-squares estimate of $r_{(s)}$ one step ahead of the present location. The predictor can use all points from the start of the task or only the last m points in computing its prediction. Alternatively, the predictor may be given a fading memory so that it discounts old information.

The averages or expectations of s and r for the nth step are given by

$$\bar{s} = \frac{1}{n} \sum_{k=1}^{n} s_n = \alpha s_n + \beta \bar{s}_{n-1} , \qquad \bar{r}_n = \frac{1}{n} \sum_{k=1}^{n} r_n = \alpha r_n + \beta \bar{r}_{n-1}$$

where

$$\alpha = \frac{1}{n} , \quad \beta = \frac{n-1}{n}$$

and

$$r_n = \begin{bmatrix} x_n \\ y_n \\ z_n \end{bmatrix} \qquad \bar{r}_n = \begin{bmatrix} \bar{x}_n \\ \bar{y}_n \\ \bar{z}_n \end{bmatrix}$$

The prediction for the next step is then computed as

$$r_{n+1} = a_n(s_n + \delta s) + b_n$$

where

$$a_n = \frac{(\overline{sr})_n - \overline{r}_n \overline{s}_n}{\overline{s_n^2} - (\overline{s}_n)^2}$$

$$b_n = \overline{r}_n - \overline{s}_n a_n$$

and

$$
\mathbf{a}_n = \begin{bmatrix} a_{xn} \\ a_{yn} \\ a_{zn} \end{bmatrix}
\qquad
\mathbf{b}_n = \begin{bmatrix} b_{xn} \\ b_{yn} \\ b_{zn} \end{bmatrix}
$$

The terms required for $\mathbf{a}_n$ and $\mathbf{b}_n$ can be recursively updated by putting them in a single vector, $\mathbf{f}_n$ such that

$$\overline{\mathbf{f}}_n = \alpha \mathbf{f}_n + \beta \overline{\mathbf{f}}_{n-1}$$

where

$$
\mathbf{f}_n = \begin{bmatrix} s_n \\ s_n^2 \\ s_n r_n \\ r_n \end{bmatrix}
\qquad
\overline{\mathbf{f}}_n = \begin{bmatrix} \overline{s}_n \\ \overline{s}_n^2 \\ (\overline{sr})_n \\ \overline{r}_n \end{bmatrix}
$$

At the nth step, new information is given the weight of $\alpha = \dfrac{1}{n}$, compared to $\beta = \dfrac{n-1}{n}$ for the previous average. As n increases, new information carries less and less weight. However, if α and b are fixed at $\dfrac{1}{m}$ and $\dfrac{m-1}{m}$ for all $n \geq m$, the new information receives a constant weight and the old information is attenuated.

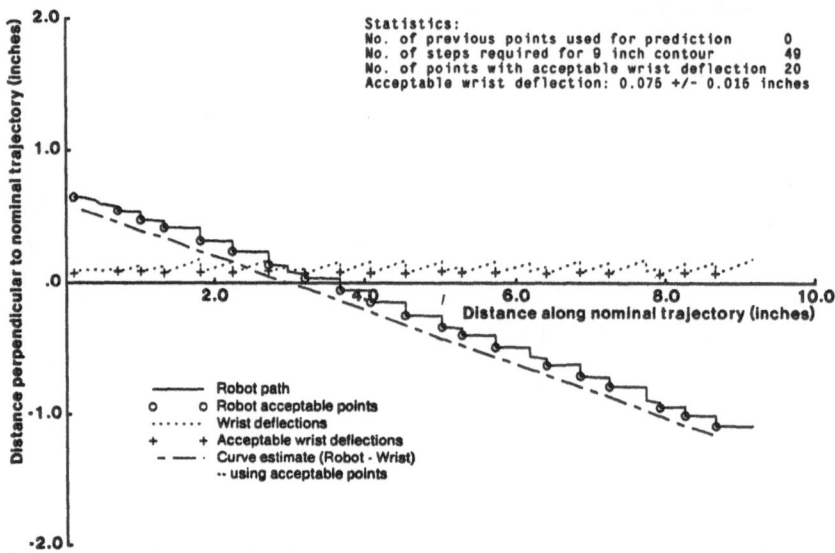

Figure 4-9: Tracking a misaligned straight contour -- no predictor

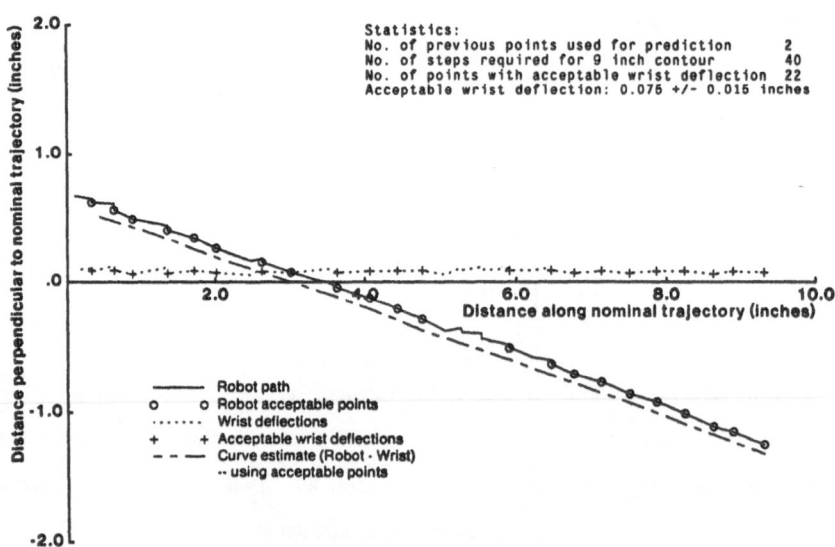

Figure 4-10: Tracking a misaligned straight contour -- two point predictor

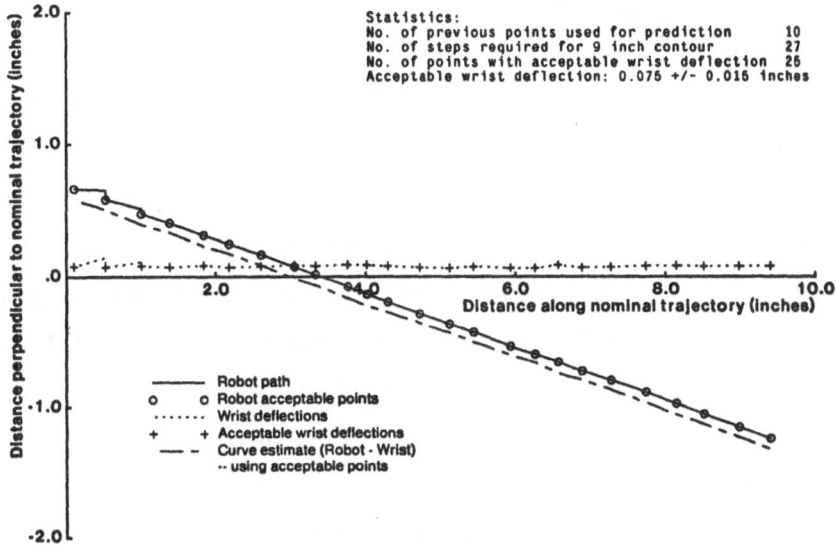

Figure 4-11: Tracking a misaligned straight contour -- ten point predictor

4.3.2 Grinding

As discussed in Chapter 3, grinding is a task in which the robot modifies the geometry of a workpiece. It is also a task in which gross and fine motions are coupled, since the velocity of the robot along the contour directly influences the grinder's depth of cut. A stiffer and more accurate robot would do a better job of grinding or routing than the general-purpose manipulator used in the current investigation. However, the purpose of the experiments was to explore the coupling between fine and gross motions in grinding and to see whether the wrist/robot combination could accomplish surface-shaping tasks. Given the flexibility and low bandwidth of the present robot and wrist, it is appropriate to take the approach that people (also with relatively flexible and low-bandwidth arms and wrists) use when they grind objects using hand tools. Therefore, a 3/4 hp electric hand grinder was used with a 4 inch fiberglass disk coated with a coarse (30 grit) aluminum oxide abrasive. The motor speed was 12,500 rpm. When people use a hand grinder, they usually keep the stiffness of their arms quite low in the direction perpendicular to the surface of the workpiece, essentially allowing the weight of the grinder to do the work. To achieve a similar effect, the grinder was mounted on a sprung, hinged bracket attached to the wrist. An additional LVDT measured the motion of the bracket with respect to the wrist. With this arrangement, the stiffness in the direction perpendicular to the surface was 20 lb/inch while the stiffness parallel to the surface was 100 lb/inch.

Specimens were fabricated with 1/16 inch tall "bumps." The robot tracked the specimens to establish the locations of the bumps and then attempted to grind them off. Unfortunately, as discussed in Section 4.2, the present robot controller does not permit continuous control of velocities from an external computer. The velocities had to be estimated, and the grinding sequence loaded into the robot as an offline program. As a result, it was not possible to achieve a smooth contour. However, the experiments did reveal the following:

- There is a minimum robot velocity of about 1 inch/second, below which the disk generates too much heat, glazes and produces burn marks on the workpiece.

- At robot velocities over 8 inches/second, with the force normal to the surface kept below 15 lb (the maximum amount that can be applied without overloading the grinder motor), the grinder removes a negligible amount of material.

- Given the above lower and upper limits on the robot velocity, it is not adequate to merely vary the velocity of the robot, while maintaining a constant force against the surface. Two additional procedures are required to remove the bumps.

  o The robot must be able to pass back and forth several times over bumps until they are removed. This method is also often used by people working with a hand grinder.

  o At least for the first pass, the robot must be allowed to press harder against the workpiece when it comes to a bump. This will make the robot inaccurate, but can be considered a "roughing" pass. The final passes can be done at the same force as used in between bumps. Again, this method is often used by people working with a hand grinder.

- The current 3/4 hp grinder is suitable for deburring and light grinding, but is too slow for removing bumps over 1/16 inch high. A more powerful grinder would require a stiffer bracket and could accept higher forces perpendicular to the surface of the workpiece.

- The inherent coupling in grinding and other shaping tasks makes it important to communicate between the wrist and the robot. For the most part the communication is one way (sending signals from the wrist to task controller), but it may also be useful to send commands to the wrist. This would be more apparent if a more powerful grinder were used. In a current application, others [109] are using an 18hp grinder with a hydraulic robot similar to the one used in the present investigation. The robot path is not modified wth sensory feedback but the velocity is reduced as the power consumption of the grinder increases. When the robot arrives at a bump, the grinding force increases 100 lb above the nominal value. If a wrist were used in such an application, it would be essential to adjust the stiffness dynamically to keep the wrist from deflecting excessively when the robot arrived at bump, while allowing it to remain sensitive when the forces were light. The same effect can be observed when people grind objects using a hand grinder. They apply more force and tense the muscles in their arms when they come to a bump that must be removed.

- At 12,500 rpm, the harmonic excitation caused by a bent or unbalanced disc is safely above the bandwidth of either the wrist or the robot. However, chatter can occur when the angle between the plane of the grinding disk and the surface is small. The solution is not to change the stiffness of the wrist, but to increase the grinding angle.

4.4 Discussion of Results

The experiments with the wrist and the robot show that a wrist/robot combination can be used for assembly, contour following and surface finishing tasks in a metal working cell.

As expected, assembly tasks can be accomplished entirely with the mechanical properties of the wrist. Surface tracking tasks require the wrist to provide filtered, sensory information to update the estimate of the surface orientation and to modify the robot path. Grinding requires that the velocity as well as the trajectory of the robot arm be modified using sensory information from the wrist. In addition, grinding may require that wrist stiffness increase as a function of robot velocity since grinding forces will increase.

The wrist, arm and task controllers form a distributed control system. The bandwidth of the overall system is limited by communications between the controllers, but this is acceptable since

- The individual elements have a degree of local intelligence and can make appropriate low-level responses to task-induced forces and displacements. In the case of the present wrist, the responses are accomplished mechanically.

- The combination of controlled compliance in the wrist and efficient, adaptive filtering and prediction algorithms permits the task controller to accurately update the robot trajectory with relatively infrequent computations (up to 1/2 inch and 1/2 second apart). The result is that a rapid feedback loop between the wrist, task controller and arm is not required.

Many of the limitations of the present system could be avoided if the wrist and arm controllers had *more* local intelligence. The amount of processing done at the task level and the amount of information passing between the task and wrist controllers could be reduced if the wrist knew how to stiffen itself in response to increased loads, or to adjust its spheres to compensate for static bending loads. If the arm controller could be told to move between points in a quadratic or cubic spline trajectory, the points could be further apart in space and time. The result would be higher tracking velocities and longer steps for a given communications bandwidth. If the arm controller also accepted commands for adjusting the servo stiffnesses and for smoothly varying the endpoint velocity, rough grinding could be satisfactorily performed.

CHAPTER 5
Analysis for an Active Robot Hand

5.1 The Promise of Further Dexterity

For some manufacturing tasks, an arm and wrist provide sufficient dexterity while a passive gripper holds the object. When human beings insert large metal parts into fixtures, or use power tools such as grinders and air wrenches, they wrap their hands around the part or tool and make fine motions with their wrists. However, for more delicate manufacturing tasks a hand with actively controlled fingers may be needed. Advantages of an active hand include

- **More versatility for fine motions**
An active hand permits the robot to attempt the tiny movements that people make in threading a bolt into a tapped hole or when using a small screwdriver to turn a screw. This is particularly true when the axis of the grasped object does not coincide with the central axis of the wrist. Since the wrist is separated by some distance from the object, it must both rotate and translate to produce a tilting (pitch or yaw) rotation about the tip of the object.[5] In fact, when the turning force on a screwdriver becomes too large to apply with fingers, people change their grip so that the screwdriver handle points along the central axis of the wrist, permitting them to twist the screwdriver without translational motions.

- **Sensors in direct contact**
With sensors in intimate proximity to the object and the task, an active hand is particularly sensitive to changes in loads upon the object and can rapidly adjust the gripping force and rigidity as required.

- **More complete loading information**
The sensors in a wrist can only determine the resultant force and torque at a point. Forces measured at the fingers, touching the object in several places, provide additional information about the distribution of the loads on the object.

---

[5]Figure 7-1 shows directions of roll, pitch and yaw manipulations with the fingers.

- **Higher possible bandwidth**
  Since there are no intermediate masses between the hand and the part, the control system of the hand can theoretically have a higher bandwidth than the arm or wrist. Salisbury [78] makes the related observation that "the proximity of low mass, powered joints to the objects being manipulated reduces the modeling errors and dynamic complexity."

- **Control of gripping forces**
  An active hand can control the gripping force, keeping it as low as possible without allowing the object to slip. This prevents damage to fragile objects and makes the robot safer in operations involving contact between the grasped object and external fixtures. It is often *desirable* for the part to slip out of the gripper when a large, unexpected force occurs (as in an assembly error, or when trying to pull an object that has become stuck).

5.2 Introduction to Grasp Analysis

In the following sections a procedure is given for determining mechanical properties with which a grip may be described. In the analysis, the arrangement of the fingers upon the object, and the stiffness and kinematic design of each finger are assumed to be known. The object is given arbitrary small displacements and the resulting motions and changes in forces are computed. From these, the overall stiffness of the grip, the ability of the grip to resist slipping and the ability of the grip to recover equilibrium in the presence of disturbances may be established. The procedure is initially illustrated with some two-dimensional examples. It is shown that the results may contain not only stiffness terms of the kind discussed by Salisbury [78] but also terms due to differential changes in the grip geometry. Unlike the stiffness terms, the geometric terms may make the grip *unstable*. A concept of grip stability which includes friction is then developed. A robot may choose between competing grips by selecting one which is stable in the presence of disturbances, which is most able to resist slipping and which matches the stiffnesses of the fingers to the compliance requirements of the task.

The analysis is extended to three-dimensional examples and explicit consideration is paid to the importance of the interaction between the fingertips and the object. Different contact conditions involving pointed, curved, soft and hard fingertips are modeled. A summary of the contact types is shown in Table 5-9. The point-contact model used in earlier analyses sometimes gives misleading results, especially when the object is small compared to the hand and when compliant gripping surfaces are employed. Finally, the results of the analysis are discussed in terms of designing and controlling dextrous hands or grippers. The results suggest that certain kinds of sensory information will be especially useful for grasp control and that a number of grasping "rules of thumb" may be argued on mechanical grounds. For example, an argument can be made for gripping as gently as possible without letting the object slip. A gentle grip not only helps to prevent damage to the object and fingers, but also (for a given combination of finger stiffnesses) results in a grip that is more likely to be stable.

Nomenclature for Grasp Properties

f_i = scalar magnitude of force applied by the ith finger

$\alpha_i \mu_i$ = acting coefficient of friction at the ith finger ($0 \leq \alpha_i \leq 1$)

μ_i = coefficient of friction at the ith finger (from surface properties)

$\mathbf{n}_i$ = unit normal vector at the ith finger

$\mathbf{l}_i$ = unit vector tangential to the object at the ith finger

$\mathbf{r}_i$ = vector from origin fixed in the object to the ith finger

$\mathbf{f}_e$ = external force taken at the origin

$\mathbf{m}_e$ = external moment taken at the origin

δn_i = normal component of displacement of ith finger

δl_i = tangential displacement of ith finger

β_i = angle between unit normal and $\mathbf{r}_i$ for i^{th} finger

k_{ni} = normal stiffness of ith finger

k_{li} = tangential or lateral stiffness of ith finger

$\mathbf{q}$ = a unit vector in an arbitrary direction

θ_q = angle between $\mathbf{q}$ and the x axis.

δq = small displacement of the object in the $\mathbf{q}$ direction

$\delta \theta$ = small rotation of the object

k_q = translational stiffness of the grip in direction $\mathbf{q}$

k_θ = rotational stiffness of the grip

$f_{\theta r}$ = restoring torque due to finger stiffnesses

$f_{\theta g}$ = grasp torque due to rotation

(*see also* Figure 5-2)

In this section the concepts of grip stiffness, strength and stability are discussed and the general procedure is described for determining the force/displacement characteristics of a grip. The concepts are illustrated at the end of the section with some simple, two-dimensional examples.

5.2.1 Grasping Model and Assumptions

A gripper may be modeled as a device with several fingers in contact with an object. The "fingers" need not resemble human fingers. They may be contact points on the jaws of a standard commercial gripper. If, for the moment, we adopt the Coulomb model of friction the static equilibrium equations become:

$$\mathbf{f}_e = \sum_{i=1}^{m} f_i(-\mathbf{n}_i) + \alpha_i \mu_i f_i(\mathbf{l}_i) \quad = \quad \sum_{i=1}^{m} f_i(-\mathbf{n}_i + \alpha_i \mu_i \mathbf{l}_i)$$

$$\mathbf{m}_e = \sum_{i=1}^{m} \mathbf{r}_i \times f_i(-\mathbf{n}_i) + \mathbf{r}_i \times \alpha_i \mu_i f_i(\mathbf{l}_i) \quad = \quad \sum_{i=1}^{m} f_i [\mathbf{n}_i \times \mathbf{r}_i + \alpha_i \mu_i(\mathbf{r}_i \times \mathbf{l}_i)]$$

(*see* **Nomenclature** *and Fig 5-1 for explanation of terms*)

The problem described by the above equations is in general statically indeterminate. The values f_i and $\alpha_i \mu_i l_i$ are the unknowns. In the above equations α_i is taken as a variable parameter between 0 and 1, so that $0 \leq \alpha_i \mu_i \leq \mu_i$, where μ_i is the standard coefficient of friction determined from surface properties. The unit vector l_i is tangential to the surface of the body but its direction is otherwise unspecified. Until the object starts to slip with respect to the fingers, l_i and α_i cannot be further defined. We can require that the above equations have at least one solution such that all $\alpha_i \leq 1$ but this is not particularly useful. It eliminates absurd finger arrangements (eg. all fingers on the same side of the object).

The presence of friction means that there are generally many grasps that will satisfy static equilibrium and it is possible to choose between them to find the one best suited for a given task. In fact, when we pick up objects with our own hands the grip we choose often depends more on what we intend to do with the object than on its shape or surface properties. For example, if we are asked to pick up a tall, thin candle that is lying on a table we may grasp it near the middle so that it balances in our hand; but if we want to push the candle into a candlestick holder we usually hold the candle near the base. Similarly, if we pick up a pen to hand it to somebody the grip we choose is entirely different from the one we use for writing.

To proceed further with a mechanical analysis it is necessary to adopt a force/deflection model for the gripper and the object. This is analogous to the use of Hookes' stress/strain relations in solid mechanics in which a model for the material provides the necessary additional equations. The force/deflection model used in the following sections incorporates a number of simplifying assumptions which are listed below.

- The fingers are modeled as elastic structures and the object as a rigid body. This is usually a good approximation for robots assembling parts or holding tools since the servoed joints in the robot arm and fingers make them considerably less stiff than the grasped object. For robots handling such materials as textiles, foamed plastic or rubber, the elasticity of the object would have to be taken into account.

- The analysis is static. There is no consideration of dynamic terms and no explicit treatment of slipping motion. However, the model can predict when a finger will start to slip upon the object and different grips may be compared by finding the one which will resist the largest task-related force or torque before slipping occurs.

- The analysis does not attempt to solve for the optimum grip for a given task but provides a mechanism for evaluating mechanical properties such as the stiffness, stability and resistance to slipping of a grip. Competing grips may be compared on the basis of such properties.

- The analysis is not concerned with geometric constraints, such as whether a gripper is actually able to achieve a given grip, or whether it is possible to place the fingers underneath an object that will be picked up from a flat surface. These are important considerations and a number of them are addressed in [110, 111, 18], but they are beyond the scope of this analysis. Basically, it is assumed that the grips under consideration have already met such criteria.

- The analysis is concerned only with small motions about an initial position. The small-motion assumption permits linear force and displacement transformations. The results of the analysis are invalid if the fingers make large motions with respect to the object, for example if they are used to turn a nut onto a bolt or to flip an object over in the hand. However, there are many tasks in which a grip is chosen and then the fingers make small motions with respect to the object. When tools such as wrenches or screwdrivers are used, the fingers usually make small motions with respect to the tool, while larger motions are accomplished with the wrist. As another example, when assembling parts, an initial grasp is chosen and then the fingers make small adjustments as the mating parts are slid together.

- Only motions with respect to the hand are considered. The interaction of the hand and the robot arm is not considered. This is not a severe restriction, however, since the compliance (inverse of stiffness) of the arm can always be added to the compliance of the hand when determining the overall force/deflection characteristics. For small and relatively low speed movements of the fingers there is little concern that dynamic coupling between hand and the arm will cause difficulties.

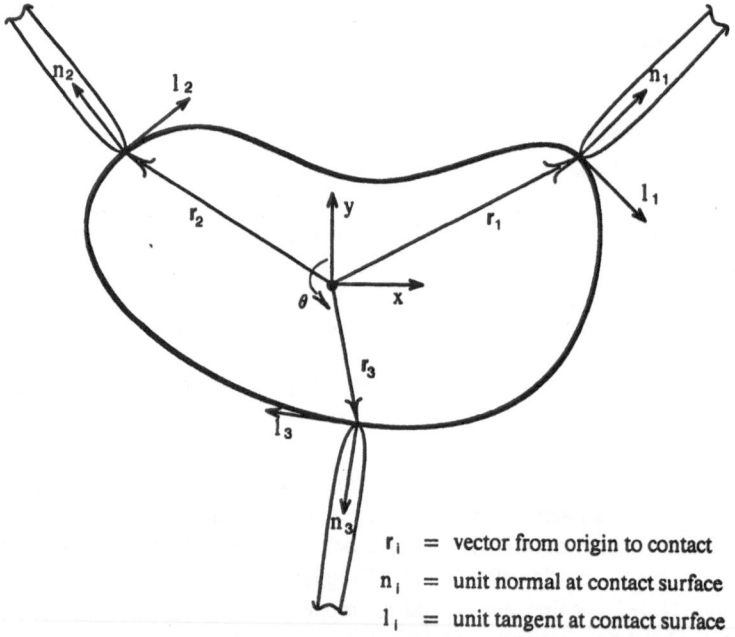

r_i = vector from origin to contact

n_i = unit normal at contact surface

l_i = unit tangent at contact surface

Figure 5-1: A two dimensional object held by three fingers

5.2.2 Stiffness, Strength and Stability of a Grasp

Stiffness

The first criterion that might be considered for evaluating a grip is the stiffness of the grip in response to externally imposed loads. The grip stiffness is a function of the stiffnesses of the fingers and of their arrangement about the object. Given a variety of possible grips, it may be useful to find the one that is stiffest with respect to torsional or translational loads. A stiff grip is useful when manipulating objects at high speeds. It helps to ensure that the displacements caused by inertial forces and torques will be small and that the natural frequency or bandwidth of the gripper/object ensemble will be high.

Robots moving freely in space are generally position-servoed and under these conditions the stiffest grip is often the best, but when a robot interacts with other objects, as during an assembly task, it becomes useful to control the mechanical *impedance* of the arm and the grip [112, 36]. Impedance control is especially well suited to servoing the fingers of the gripper or hand [19]. At low speeds, dynamic effects become negligible and impedance control reduces to stiffness control. For example, the robot hand can be made stiff in directions which are not constrained by contact with fixtures and compliant in the directions which are. In terms of choosing a grip, the best grip is the one which best matches the requirements of the task to the achievable range of finger stiffnesses.

Resistance to slipping

A second way to discriminate between grips is to find the one that, for a given combination of servo stiffnesses, grasping forces and fingertip geometries, can resist the greatest possible applied force or torque before any of the fingers slip. This again is desirable when manipulating objects at high speed. For tasks involving contact forces and torques the same analysis may be used to find the grip for which the fingers are least likely to slip in response to the expected range of forces and torques.

Stability

A third criterion is grip stability. Since the analysis is linearized and only small motions are considered it is only possible to determine whether a grip is *infinitesimally stable*, that is, whether the grip will return to its original position if the object is displaced by an arbitrary small amount. This amounts to determining whether the changes in the forces on the object that result from disturbing it will tend to oppose or to increase the disturbance.

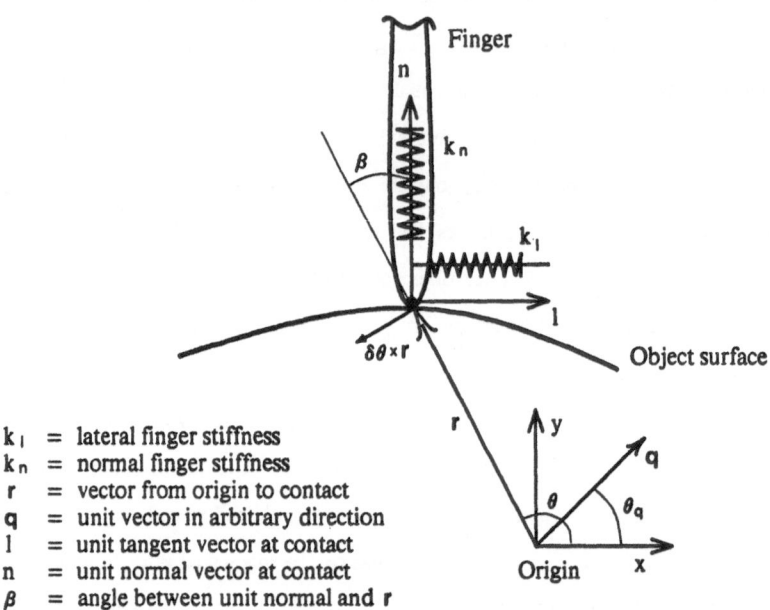

k_l = lateral finger stiffness
k_n = normal finger stiffness
r = vector from origin to contact
q = unit vector in arbitrary direction
l = unit tangent vector at contact
n = unit normal vector at contact
β = angle between unit normal and r

Figure 5-2: Detail of a single two-dimensional finger from Figure 5-1

5.2.3 Procedure for Establishing Grip Properties

The procedure used in determining the above grip properties is outlined below.

1. Displace the object an arbitrary, small amount.

2. Determine the resulting motions of the fingers. These will depend on the finger geometries, contact types and stiffnesses.

3. Determine the changes in the forces at the finger/object contact areas that result from the motions of the object and fingers. There are two contributors to these changes. The first are restoring forces that result from the stiffnesses of the fingers. The second result from changes in the grip. The fingers and the object do not move together as a rigid ensemble and the resulting modification of the grip geometry changes the way in which the finger forces act upon the object.

4. Compare the new forces at the finger/object contact areas with the maximum forces that the contacts can sustain without slipping. Also determine whether the normal forces would become negative at any of the fingers (meaning that they would lose contact with the object).

5. Compare the new resultant forces and torques on the object with the original forces and torques and with the displacement of the object to determine the stiffness and infinitesimal stability of the grip.

In later sections, particular attention is paid to the interactions between different kinds of fingertips and an object. Curved, soft, and pointed fingertips are discussed and their effects on the grip are investigated. It is shown that the point-contact model adopted in earlier analyses is only accurate when the fingertips are small compared to the object being held. Thus, if we hold a large cardboard box or a basketball, a point-contact model of our fingertips is fairly accurate, but when we hold a matchbox or a golf ball it is not.

5.2.4 Two-Dimensional Examples

The concepts of grip stiffness, stability and resistance to slipping can be illustrated with some short examples. In these two-dimensional examples, the forces and motions are broken into scalar components, but a matrix notation will be used for the three-dimensional analysis in later sections. Figure 5-1 shows a rigid body held by three fingers which are assumed to have some characteristic stiffness. The actual stiffness of each finger need not be prescribed; only the *relative* stiffness with respect to the other fingers is required. In the following two-dimensional examples, the finger stiffnesses may be resolved into components k_{ni} and k_{li}, perpendicular and parallel to the surface of the object. As before, the fingers need not resemble human fingers but may be the contact areas of an industrial gripper. It is required only that their stiffness and friction characteristics be known. Figure 5-2 shows the coordinates and stiffnesses for a single finger.

Looking first at torsional loading, if a force is externally applied to the object, (perhaps by a wrench at the x,y origin in Figure 5-1), the object will be rotated by a small amount, $\delta\theta$. Each fingertip in contact with the object must move $\delta\theta\times\mathbf{r}_i$ along with the object surface. The finger motions can be resolved into components parallel and perpendicular to the surface of the object.

$$\delta n_i = (\delta\theta \times r_i) \cdot n_i = -r_i \delta\theta \sin\beta_i$$

$$\delta n_i = (\delta\theta \times r_i) \cdot l_i = -r_i \delta\theta \cos\beta_i$$

We can equate the potential energy stored in rotating the body with the energy stored in the fingers to express the rotational stiffness of the grip in terms of the finger stiffnesses.

$$\tfrac{1}{2} k_\theta \, \delta\theta^2 \quad = \quad \sum_{i=1}^{m} \tfrac{1}{2} k_{ni} \delta n_i^2 + \tfrac{1}{2} k_{li} \delta l_i^2$$

Substituting for δn_i and δl_i,

$$k_\theta \quad = \quad \sum_{i=1}^{m} r_i^2 (k_{ni} \sin^2\beta_i + k_{li} \cos^2\beta_i)$$

The stiffest grip for torsional loading is that for which k_θ is greatest.

To find the grip that will resist the greatest torsional load without slipping we first look at each of the grips under consideration and discover which finger (or fingers in a symmetrical grip) is nearest to slipping for a given applied moment, m_e, at the origin. As the body is rotated $\delta\theta$, the changes in the forces at each finger are

$$\delta f_{ni} = k_{ni} \, \delta n_i \quad \text{and} \quad \delta f_{li} = k_{li} \, \delta l_i.$$

From the discussion earlier in this section, $f_{li} = \alpha_i \mu_i f_{ni}$ for the Coulomb law of friction, where slipping will occur as $\alpha \to 1$. Then, for example, if initially $f_{li} = 0$, slipping will occur when

$$k_{li} \, \delta l_i > \mu_i f_{ni}$$

Thus, for a given rotation, $\delta\theta$, the finger nearest to slipping will be the one for which α is closest to 1, or for which

$$\alpha_i = \frac{k_{li} \delta l_i}{\mu_i f_{ni}} \quad = \quad \frac{m_e k_{li}(-r_i \cos\beta_i)}{k_\theta \mu_i f_{ni}}$$

is greatest.

Having found the "worst case" finger for each grip we chose between grips by finding the one for which m_e is greatest before $\alpha = 1$ at the finger.

$$m_{e\,max} \quad = \quad -\frac{k_\theta \mu_j f_{nj}}{k_{lj} r_j \cos\beta_j}$$

(where j is the subscript of the "worst case" finger)

For motion in an arbitrary direction, q, the angle at each finger between q and n_i is $\varphi_i = \theta_i - \beta_i - \theta_q$, (where q, n, θ, β and θ_q are shown in Figure 5-2 for a typical finger). Equating potential energies allows the translational stiffness to be expressed in terms of the finger stiffnesses.

$$k_q = \sum_{i=1}^{m} k_{ni} \cos^2\varphi_i + k_{li} \sin^2\varphi_i$$

Following the procedure used for the rotational case, we can choose between grips to find the one that will withstand the largest force, f_e, in a given direction, q, before any of the fingers slip. The "worst case" finger is the one for which

$$\alpha_i = \frac{k_{li}\delta l_i}{\mu_i f_{ni}} = \frac{f_e k_{li} \sin\varphi_i}{k_q \mu_i f_{ni}}$$

is greatest. The best grip is then the one for which f_e can be greatest before the "worst case" finger will slip.

$$f_{emax} = \frac{k_q \mu_i f_{nj}}{k_{lj} \sin\varphi_j}$$

5.2.4.1 Choosing among five grips: an example

Figure 5-3 shows five grips on a rectangular block. Grips 1, 4 and 5 share the same configuration, but with different finger spacings. We can use the above results to discriminate between the grips. To simplify the computation we assume that the fingers are all identical and that their stiffness components, k_{ni} and k_{li}, are independent of the orientation of the finger. This is a reasonable approximation for long fingers with several joints.

The highest rotational stiffness is achieved either with grip 1 or grip 2, depending on whether k_{ni} or k_{li} is greater. If it is most important that none of the fingers slip when a moment is applied to the rectangle, then grips 1, 4 or 5 should be chosen. Grip 1 offers the best combination of rotational stiffness and resistance to slipping.

For translations the picture is a bit more complicated since the stiffness and the resistance to slipping vary as the direction of q varies. Intuitively, one might suggest that grip 3 is the safest choice. Figures 5-4 and 5-5 show plots of the stiffness and the maximum force without slipping as a function of angle, θ_q. For these plots, k_{ni} was arbitrarily taken twice as large as

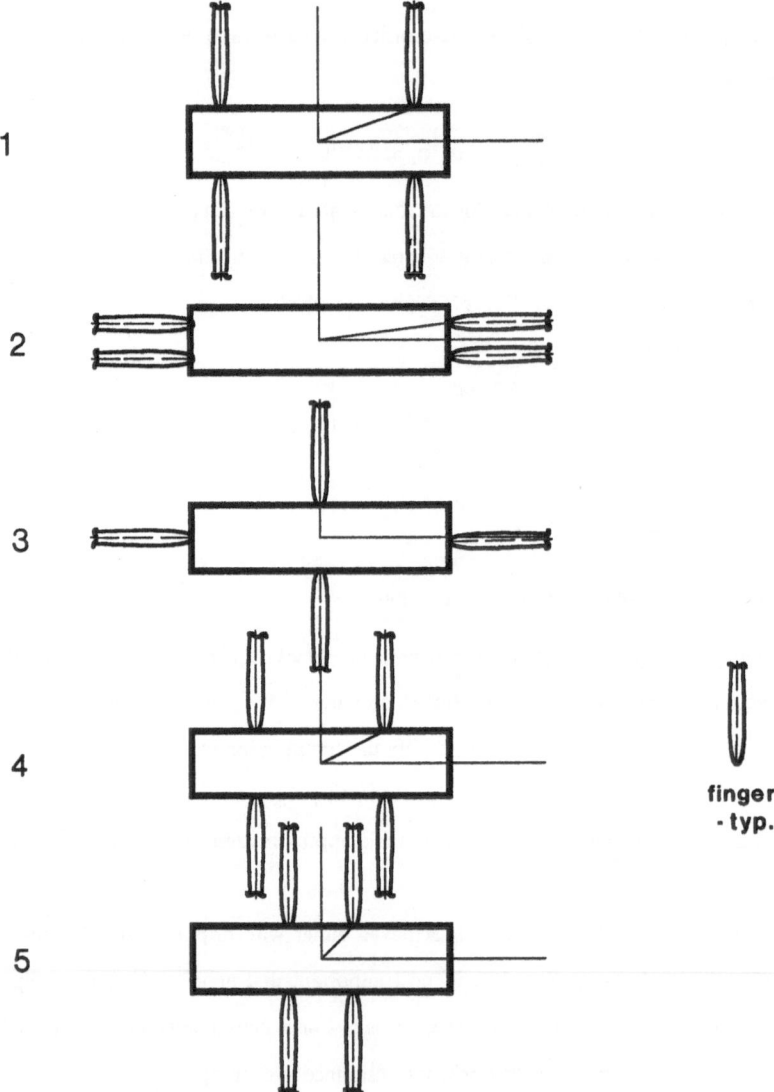

Figure 5-3: Five ways to grip a rectangle with four fingers

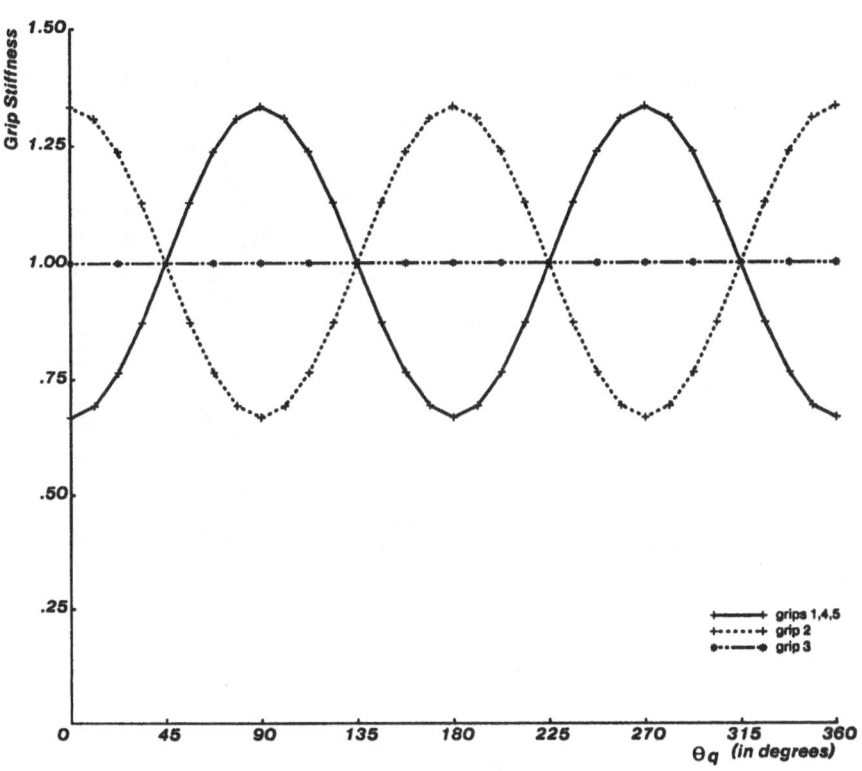

Figure 5-4: Grip stiffness for force at angle θ_q

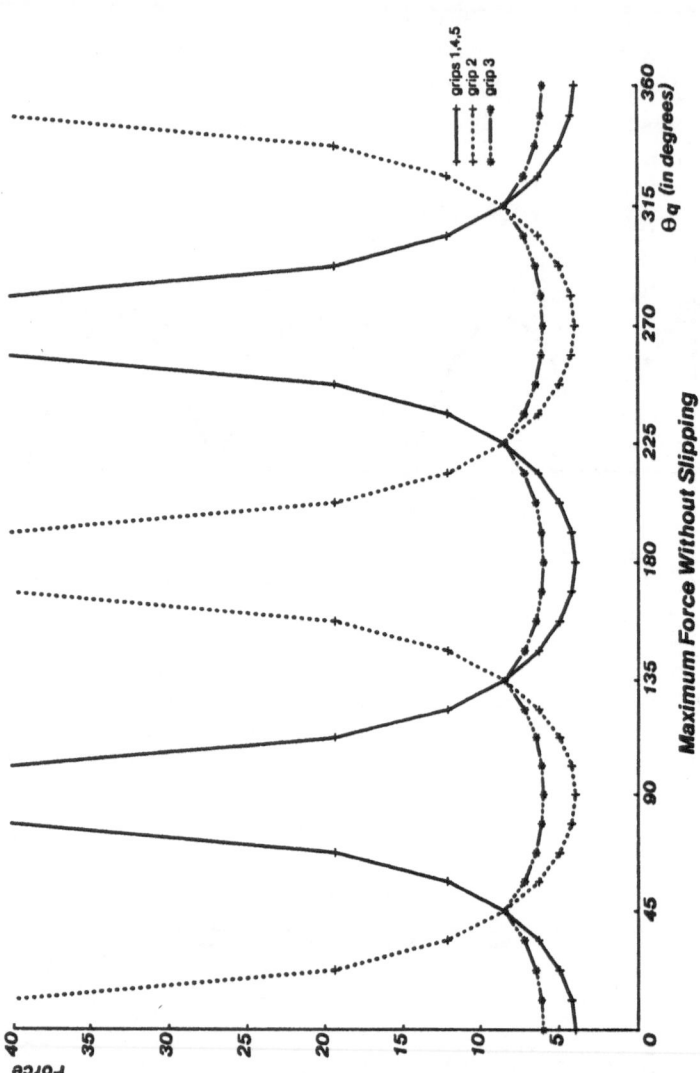

Figure 5-5: Maximum force without slipping at angle θ_q

k_{li}. Actual values of k_{ni} and k_{li} might be quite different, but the plots provide an example of how grip stiffness varies as a function grip geometry. In this case, the stiffness of grip 3 is constant, regardless of the direction of $\mathbf{f}_e$. Grip 3 also offers the most nearly constant resistance to slipping and is therefore the safest choice for arbitrary loads, although other grips offer more stiffness or resistance to slipping when the object is pulled in a single direction.

5.2.4.2 An unstable example

The foregoing discussion has focused on determining whether the stiffness of a grip is suitable and on determining when the fingers slip. The next question is whether the grasp will be stable if perturbed slightly. A potentially unstable grip is shown in Figure 5-6. If the grasp forces are large, and if the fingers are not stiff enough in the lateral direction, the rectangle will continue to rotate when disturbed by a small angle, $\delta\theta$, instead of returning to the initial position. The same effect can be seen by gripping a coin on edge between two opposed fingers. If one squeezes too hard, the coin "collapses" to a more stable position in which one's fingers are pressing against the faces. In general the coin will also slip with respect to the fingers when this occurs, but before slipping occurs it is possible to determine whether the grip is stable.

As the rectangle in Figure 5-6 is rotated by a small angle, $\delta\theta$, the lateral stiffnesses of the two fingers produce a restoring torque, $f_{\theta r}$

$$f_{\theta r} = 2k_l\, r^2\, \delta\theta$$

At the same time, due to the rotation of the body, a torque is generated by the grasp forces,

$$f_{\theta g} = 2f_n r\, \delta\theta$$

The net change in the torque upon the object is

$$\delta m_e = (f_n - k_l r)2r\delta\theta$$

The grip is unstable if the change in the torque is positive for a positive rotation, $\delta\theta$. Thus, for the grip to be infinitesimally stable it is required that $f_n < k_l r$. Evidently, for a given rectangle size and finger stiffness, pressing harder makes the grip less stable. This result appears again in later examples and provides an incentive for not gripping harder than

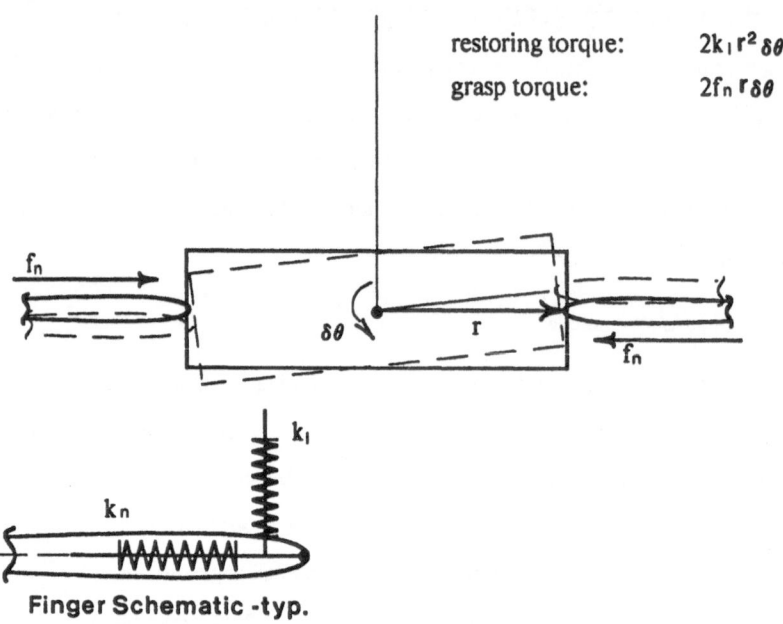

restoring torque: $2k_1 r^2 \delta\theta$

grasp torque: $2f_n r \delta\theta$

Finger Schematic -typ.

Figure 5-6: Instability of a rectangle held by two fingers

necessary because for a given grip geometry, the stability of the grip *decreases* with increased gripping force. Another result is that for a given finger stiffness and gripping force, the grip is more stable for a longer rectangle (one for which r is large).

These effects can be demonstrated by pressing a pencil lengthwise between the index fingers of each hand. As one presses harder the grip is likely to collapse unless one also tenses (stiffens) one's arm and finger muscles. If the experiment is repeated for an old, short pencil and for a new, long one it will be seen that the grip collapses more easily for the short one.

Unfortunately, if we return to the example of gripping a coin between two fingers of one hand, a problem appears. If the fingers are now pressing against the faces of the coin instead of the edges, the grip should, according to the above equation, become *less* stable. This is clearly incorrect and demonstrates that the point-contact fingertip model gives inaccurate results for human fingers pressing against the faces of a coin. If we repeat the example, using ball-point pens instead of our fingers to press against the faces of the coin, we find that the grip is indeed very unstable. The problem is resolved if we model the finite curvature and deformation of our fingertips. Thus, in the following sections, a framework is established in which examples like those above can be extended to three dimensions and in which fingers with pointed, curved and soft contacts are considered.

Nomenclature for Three-Dimensional Grasp Analysis

o $\quad\quad$ = origin of (x,y,z) system

bp $\quad\quad$ = origin of (l,m,n) system and contact point on object

fp $\quad\quad$ = contact point on fingertip

f $\quad\quad$ = origin of (a,b,c) system

r_b $\quad\quad$ = 3x1 vector from (x,y,z) origin to (l,m,n) origin

r_f $\quad\quad$ = 3x1 vector from (a,b,c) origin to (l,m,n) origin

d_b $\quad\quad$ = vector of small translations and rotations of the object in (x,y,z) coordinates

d_{bp} $\quad\quad$ = vector of small translations and rotations of the object in (l,m,n) coordinates

d_c $\quad\quad$ = vector of displacements transmitted through the contact

d_{fp} $\quad\quad$ = vector of small finger translations and rotations in (l,m,n) coordinates

d_f $\quad\quad$ = vector of small finger translations and rotations in (a,b,c) coordinates

d_q $\quad\quad$ = vector of small finger translations and rotations in joint coordinates

g_b $\quad\quad$ = vector of forces and torques on the object in (x,y,z) coordinates

g_{bp} $\quad\quad$ = vector of forces and torques on the object in (l,m,n) coordinates

g_c $\quad\quad$ = vector of forces transmitted through the contact

g_{fp} $\quad\quad$ = vector of finger forces and torques in (l,m,n) coordinates

g_f $\quad\quad$ = vector of finger forces and torques in (a,b,c) coordinates

g_q $\quad\quad$ = vector of finger forces and torques in joint coordinates

$[Jb]$ $\quad\quad$ = 6x6 jacobian relating d_b to d_{bp}

$[Jf]$ $\quad\quad$ = 6x6 jacobian relating d_f to d_{fp}

$[Jq]$ $\quad\quad$ = nfx6 jacobian relating d_q to d_f

$[Jfq]$ $\quad\quad$ = nfx6 product of $[Jf]$ and $[Jq]$

$[P]$ $\quad\quad$ = partition of $[Jfq]$

$[P^*]$ $\quad\quad$ = non-singular partition of $[Jfq]$

$[G]$ $\quad\quad$ = 9x9 grasp jacobian for three fingers

$[Kq]$ $\quad\quad$ = stiffness matrix of a finger in joint coordinates

$[Kf]$ $\quad\quad$ = stiffness matrix of a fingertip for three-fingered hand

$[Kx]$ $\quad\quad$ = $[Kf]$ rotated to world coordinates

$[Kb]$ $\quad\quad$ = stiffness matrix of the grasp

$[Cf]$ $\quad\quad$ = compliance matrix for finger and fingertip

[A]	= 3x3 orthonormal rotation matrix
[R]	= 3x3 skew-symmetric matrix for $\mathbf{r}$
[I]	= the identity matrix
[M]	= matrix of contact degrees of freedom
[L]	= square matrix assembled from [P] and [Kq]
l	= vector of Langrange multipliers for [L]
λ_i	= i^{th} Langrange multiplier
q_f	= set of nf independent elements in $\mathbf{d_{fp}}$
$\mathbf{u_f}$	= 3x1 unit tangent vector on fingertip
$\mathbf{u_b}$	= 3x1 unit tangent vector on object
nf	= number of degrees of freedom of finger
nc	= number of force or displacement components transmitted through contact
s	= arclength along fingertip or object surface
r_c	= magnitude of radius of curvature of fingertip
r_o	= outer radius of contact area
E	= modulus of elasticity
G	= shear modulus
A	= contact area
I_{ij}	= ij moment or product of inertia
I_p	= polar moment of inertia
σ_{ii}	= stress in ii direction
τ_{ij}	= ij shear stress
k	= scalar stiffness component
f	= scalar force component
w	= width
t	= thickness

5.3 Extension to Three-Dimensional Problems

5.3.1 Forward Force and Displacement Relations

In the general case, the gripper fingers and the object may have up to three translational and three rotational degrees of freedom. It becomes convenient to use matrix equations to express the grip stiffness, strength and stability. In the following discussion, force vectors, g or f, include force and moment components and displacement vectors, d, include small translation and rotation components:

$$f^t = [f_x, \ f_y, \ f_z, \ f_{\theta x}, \ f_{\theta y}, \ f_{\theta z}]$$
$$d^t = [d_x, \ d_y, \ d_z, \ d_{\theta x}, \ d_{\theta y}, \ d_{\theta z}]$$

The goal of this analysis is to express the interaction between grasping forces and small motions of the object. If g_b is the resultant grasp force on the object and d_b is a vector of small motions of the object then the desire is to determine

$$\frac{\partial g_b}{\partial d_b} = ?$$

Since d_b is a small quantity this may be approximated by the linear relationship

$$\Delta g_b = [?] d_b$$

where $[?]$ is a matrix that must be determined. To do this it is necessary to first establish how the forces applied by the fingers, g_f, determine the grasp force, g_b, and to establish the relationship between a small motion of the object, d_b, and the resulting motions of the fingers, d_f. If g_b and d_b were scalars, δf and δx, the relationship between them could be written

$$\frac{\partial f}{\partial x} = k \quad \text{or} \quad \delta f \approx k \delta x$$

Under certain circumstances, for example if the fingers do not move relative to the object when the object moves slightly, an equivalent stiffness expression can be written for forces and displacements of the object

$$g_b = [Kb] d_b$$

where $[Kb]$ is a symmetric stiffness matrix. More commonly, the fingertips and the contact areas will shift with respect to the grasped object as it moves and new terms are added to the above stiffness relationship. Such terms are discussed later in this section.

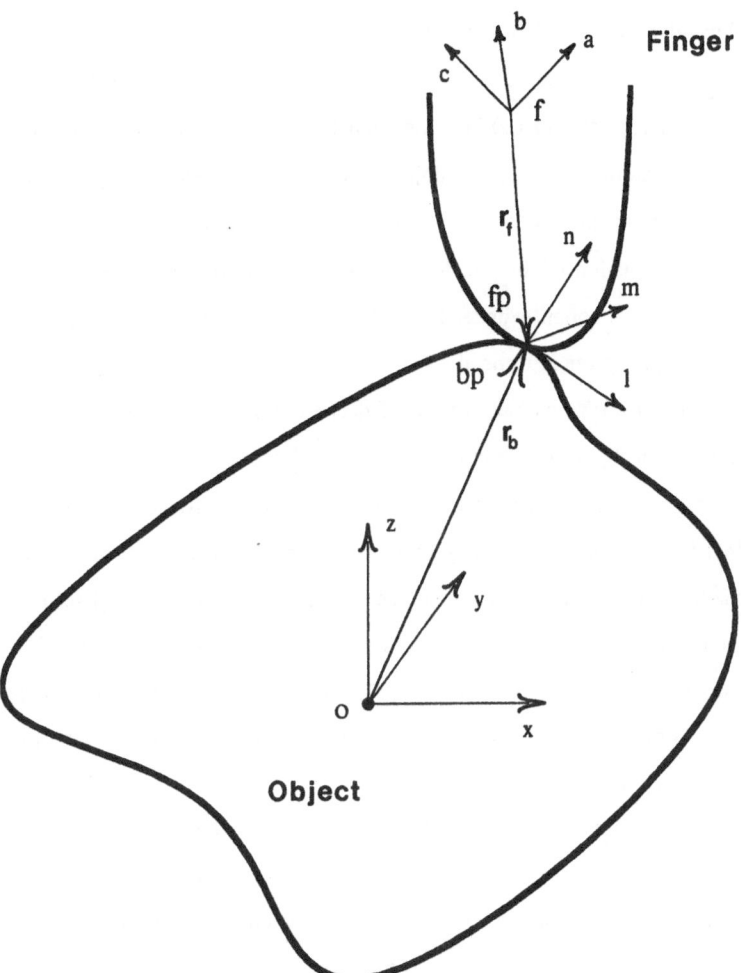

Figure 5-7: Coordinate systems for a finger touching an object

In Figure 5-7, the coordinate systems are shown for a fingertip touching an object. The fingertip may be the last segment of a multijointed finger or it may be a contact surface on the jaw of an industrial gripper. The global coordinate system, (x,y,z), is embedded in the object at o. The (a,b,c) coordinate system is embedded in the fingertip at f and, like the (x,y,z) system, may be chosen with any convenient position and orientation. The (l,m,n) coordinate system is shared by the fingertip contact area, fp, and the object contact area, bp. The n axis is perpendicular to both the object and fingertip surfaces and the l,m axes lie in the common tangent plane. The finger joint coordinates are not shown in Figure 5-7 since they will be different for each finger design.

It is assumed that the position and orientation of the (a,b,c) coordinate system can be determined with respect to (x,y,z) from the geometry of the gripper and knowledge about the initial position and orientation of the object. Salisbury [78] has shown that the position and orientation of the tip of a multijointed finger may be established in the same way that the position and orientation of the end link of a manipulator are determined from the joint angles. The result is often expressed as a 4x4 transformation matrix, $[T]$, [113]. The elements of $[T]$ are given in Appendix A.1.

Usually, the fingertip will have less than 6 degrees of freedom and the compliance of the fingertip will be negligible in one or more directions. For example, a finger with $nf < 6$ joints is often considered to have nf degrees of freedom since the structural compliance of the finger links is negligible in comparison to the compliance of the servoed joints. In this case, the displacement vector of the finger in joint coordinates, d_q, will be an nfx1 vector.

The fingertip is also assumed to have known stiffness properties, represented by the nfxnf stiffness matrix $[Kq]$ in joint coordinates. Salisbury [78] has shown that the stiffness matrix for the tip of a multi-jointed finger, valid for small motions, may be derived from the finger kinematics and joint servo gains.

Frequently, the fingertip may be treated as a rigid body so that small displacements of the finger in joint coordinates may be related to displacements in the (a,b,c), which in turn, may be related to displacements in the (l,m,n) system with the linear transformations:

$$d_f = [Jq] d_q \quad \text{(where } [Jq] = \frac{\partial d_f}{\partial d_q} \quad \text{defines a Jacobian)} \tag{5.1}$$

$$d_{fp} = [Jf] d_f \tag{5.2}$$

$$d_{fp} = [Jfq] d_q \quad \text{where } [Jfq] = [Jf][Jq] \tag{5.3}$$
$$\phantom{d_{fp} = [Jfq] d_q \quad} (nf\text{x}6) \quad (6\text{x}6)\,(nf\text{x}6)$$

The fingertip displacement vector, d_{fp}, will contain 6 elements of which nf will be linearly independent. A set of nf linearly independent elements within d_{fp} is called q_f.

The object is treated as a rigid body and consequently, a small motion, d_b, of the object in the (x,y,z) system produces a displacement of the contact area, d_{bp}, in the (l,m,n) system.

$$d_{bp} = [Jb] \, d_b \qquad\qquad (5.4)$$

For generality, d_b and d_{bp} are taken as 6 element vectors (possibly with some zero elements). A number of identities for 6x6 Jacobians are given in Appendix A.1.

It can be shown [113], by equating virtual work, that small displacements and forces transform in a complementary way. If the grasping force, g_{fp}, at the fingertip contact area is known then the equivalent force in the (a,b,c) system is found by equating the work done in displacing the fingertip by d_{fp} and the finger by d_f.

$$d_{fp}{}^t \cdot g_{fp} = d_f{}^t \cdot g_f$$

Then, substituting from equation (5.2),

$$d_f{}^t \cdot g_f = d_f{}^t [Jf]^t \, g_{fp} \quad \text{or} \quad g_f = [Jf]^t \, g_{fp} \qquad\qquad (5.5)$$

Similarly,

$$g_q = [Jq]^t \, g_f \qquad\qquad (5.6)$$

and

$$g_b = [Jb]^t \, g_{bp}. \qquad\qquad (5.7)$$

5.3.2 Summary of Forward Transformations

The forward displacement and force transformations are summarized in Figure 5-8.

Starting at the lower left corner with a displacement, d_b, of the object in (x,y,z) coordinates, and following the arrows, the displacements transmitted through the contact are determined as d_c. Then, starting with the contact forces, g_c, on the object in (l,m,n) coordinates, and following the arrows, one computes the forces upon the object

Starting at the lower right corner with displacements of the finger joints, d_q, the displacement of the fingertip, d_{fp} ca be determined. Finally, if the contact forces, g_{fp}, are known for the finger, following the arrows gives the forces in the finger joints, g_q.

The forward *displacement* relations provide transformations starting with the object or the finger and working towards the common contact. The forward *force* relations start with

Object force   **Finger joint torques**

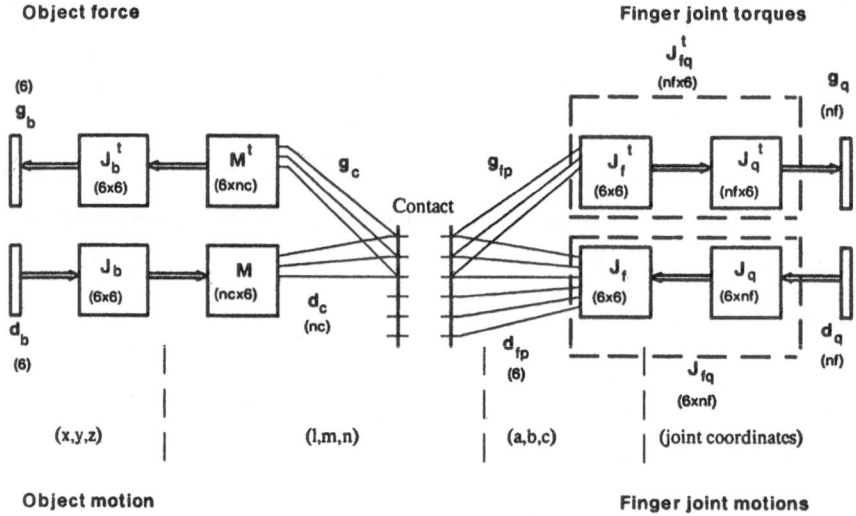

Figure 5-8: Flow chart for forward force and displacement transformations

the contact and work outward toward the object or the finger. Unfortunately, these relations are not sufficient to complete steps 2 and 3 of the procedure outlined in Section 5.2.3. Once d_c has been determined, an *inverse* relationship giving d_q in terms of d_c must be used. The solution depends on the type of contact and the number of degrees of freedom of the finger, and is discussed in Section 5.3.3. Once d_q has been determined, another inverse relation is required to determine the change in g_{fp}. This solution also depends on the contact and the finger, and is discussed in Section 5.3.5. A forward force transformation can then be used to determine the change in g_b from the change in g_{fp}.

5.3.3 Finger Motions and Constraints

The mobility of an object represents the number of degrees of freedom with which the object can make arbitrary motions. The mobility is subject to constraints imposed at each contact point which may prevent motions in certain directions and couple the motions of the object in others. Generally, the mobility of the object decreases as the number of contact points increases.

The determination of mobility involves first finding the constraints imposed at each contact point and then determining how the different contacts interact to limit the mobility of the object. In this section, the emphasis is on characterizing the constraints and contact conditions for a single contact so that various fingertips may be compared.

The way in which the constraints at each finger combine to constrain an object is discussed in previous analyses [78, 86]. For such analyses it is convenient to adopt the terminology of *wrenches* and *twists* in which the magnitudes of the components of the force and displacement vectors, g_{bp} and d_{bp}, are separated from their directions. The number of degrees of freedom of the object depends on the *intersection* over all contacts of the degrees of freedom from each [86]. The number of independent forces that may be applied to the object by the hand increases as the *union* over all contacts of the forces that each can apply. When more than one contact can apply forces in the same directions, it becomes possible to specify internal forces on the object [78, 86]. These may be set to ensure that all fingers remain in contact with the object.

5.3.4 Constraints at a Contact

At each contact point, the constraints depend on how many degrees of freedom are transmitted through the contact and on how many degrees of freedom the finger has. Basically, there are three categories. In the first case a motion of the object *exactly determines* the motion of the finger (this is the simplest case, in which a part of $[Jfq]$ is simply inverted for the inverse displacement and force relations). In the second case the motion of the finger is *under determined* and in the third case the motion of the finger is *over determined.*

Forces and motions at the fingertip-object contact area are transmitted through a coupling matrix, $[M]$. The elements of $[M]$ depend on the contact geometry (*see* Figure 5-12) and friction conditions. These are discussed further in Section 5.4. If there is complete coupling in six degrees of freedom between the object and the fingertip (as in the case of a soft, sticky finger adhering to the object) $[M]$ becomes a 6x6 identity matrix. The elements of d_{fp} that are transmitted to the finger form the vector d_c and the elements of the grasp force, g_{fp}, that are transmitted to the object form the vector g_c which has *nc* components.

$$d_c = [M]d_{bp} \qquad g_c = [M]^t g_{fp} \qquad (5.8)$$

The contact constraints are found by comparing the elements of d_c with the independent members of d_{fp}. As mentioned in the last section, nf elements of d_{fp} will usually be linearly independent for a finger with nf joints. A set of nf independent elements within d_{fp} is called q_f and, for the purposes of describing the contact constraint, there are three conditions:

1. A set of independent elements in d_{fp} can be found such that $d_c = q_f$ and $nc = nf$. In this case arbitrary motions of the object at bp are possible and the motion of the fingertip is completely determined. Similarly, the joint torques of the finger completely determine the set of forces, g_c, that can be transmitted through the contact to the object.

2. A set of independent elements in d_{fp} can be found such that $d_c \subset q_f$. If d_c is a subset of q_f, arbitrary motions of the object at bp are possible but the finger motion is not completely determined. The remaining undetermined elements of d_{fp} or d_q may be solved for by requiring that the finger move so as to minimize its potential energy.

3. $d_c \not\subset q_f$. If d_c contains elements that are not included in q_f, the finger and contact limit the possible motions of the object. At the same time, it is possible that $q_f \not\subset d_c$, in which case a (constrained) motion of the object does not completely determine d_{fp}. If this happens, the undetermined elements of d_{fp} must be determined as above.

Methods for solving for the motions of the finger are discussed below for each of the above situations. In each case, a submatrix, $[P]$, is extracted from $[Jfq]$ that relates the nc elements of d_c to the nf elements of d_q: $d_c = [P]d_q$. The three cases are identified by evaluating the rank of $[P]$.

5.3.4.1 Case 1: exactly determined

An example for which $d_c = q_f$ and $nc = nf$ is a finger with three joints, constructed so that the fingertip can move in three directions, always touching the object at a single point fixed on the object surface. This is mathematically the most convenient situation and forms the basis of previous investigations on grip stiffness [78]. The matrix, $[P]$, that is extracted from $[Jfq]$ will be square and non-singular. The relations are:

$$\text{rank}([P]) = nf = nc$$

$$d_q = [P]^{-1}d_c \tag{5.9}$$

5.3.4.2 Case 2: under determined

When $d_c \subset q_f$, the submatrix, $[P]$, that relates the nc members of d_c to the nf joint variables, d_q, will have rank nc. The motion of the fingertip will minimize the potential energy of the finger, subject to the nc constraint conditions that make up the rows of $[P]$. The change in the potential energy of the finger may be expressed as

$$\Delta P.E. = g_q{}^t d_q + \tfrac{1}{2}(d_q{}^t [Kq] d_q) \tag{5.10}$$

in which the first term is due to work done against the grasping joint torques and the second is due to the stiffnesses of the finger joints. The second term is what provides the grasp stiffness discussed in previous investigations [78, 77], but the first term may be of comparable magnitude.

To minimize the potential energy, the magnitude of the above expression must be at a maximum. If the elements of d_q were all independent (i.e. if there were no coupling between d_c and d_q) then the maximum would be found by taking the partial derivative of the above equation with respect to each member of d_q and setting the resulting expressions equal to zero. In the present case, a flexible and systematic approach is to use Lagrange multipliers. The resulting equation is conveniently expressed as

$$\begin{array}{|c|}\hline d_q \\ \hline l \\ \hline \end{array} = [L]^{-1} \begin{array}{|c|}\hline g_q \\ \hline d_c \\ \hline \end{array}$$

where $[L]$ can be assembled from $[P]$ and $[kq]$ and l is a vector of Langrange multipliers. Details are given in Appendix A.2.

Once all the members of d_q have been found, the motion of the finger in (l,m,n) coordinates is found using $d_{fp} = [Jfq] d_q$. The restoring forces in the joints are given by $\Delta g_q = [Kq] d_q$. Since $[P]$ is not square, $[P]^{-t}$ cannot be used as in Case 1 to determine the changes in the forces at the fingertip, δg_{fp}. However, since $d_c \subset q_f$ and since d_q have been determined subject to the constraints of $[P]$, some columns may be removed from $[P]$ so as to leave a square matrix, $[P^*]$ relating d_c to nc of the nf elements in d_q.

5.3.4.3 Case 3: over determined

When $d_c \not\subset q_f$ the elements of d_c become coupled and the object is constrained by the finger and contact. The submatrix, $[P]$, will have a rank of less than nc. In this case, rows of $[P]$ corresponding to particular elements of d_c may be eliminated to produce a smaller matrix, $[P^*]$ that has the same rank as $[P]$. The elements of d_c corresponding to $[P^*]$ form the vector, d_c^*. If the new submatrix, $[P^*]$ has rank nf then it may be inverted as in Case 1 to determine d_q from d_c^*. All the elements of d_{fp} can then be recovered as $[Jfq]d_q$. Thus, the kinematic coupling between the elements of d_{fp} is defined.

If the rank of $[P]$ is less than nf then the motion of the finger is not completely determined and potential energy methods must be used as in Case 2 to determine d_q from d_c^*. Again, the complete motion of the fingertip is recovered from $[Jfq]d_q$.

The general method for determining the motions of a finger from the motions of the contact is illustrated in the left hand portion of Figure 5-10. For the particular case in which $[P]$ is invertible, the method is summarized in Figure 5-9.

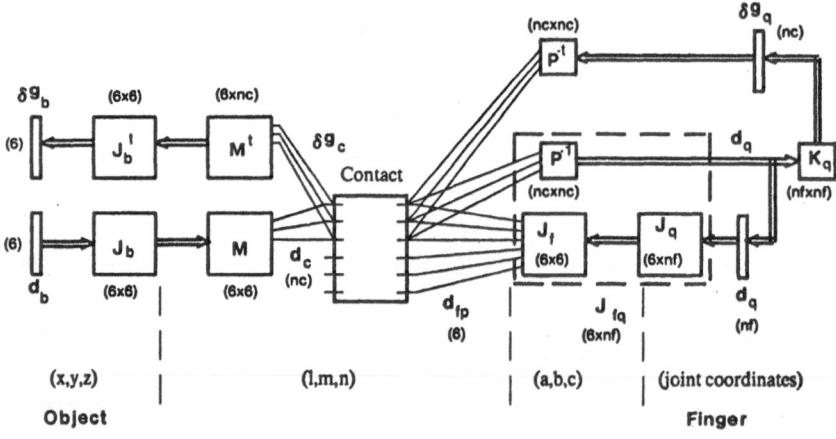

Figure 5-9: Flow chart for cases in which $[P]$ is invertible (Case 1)

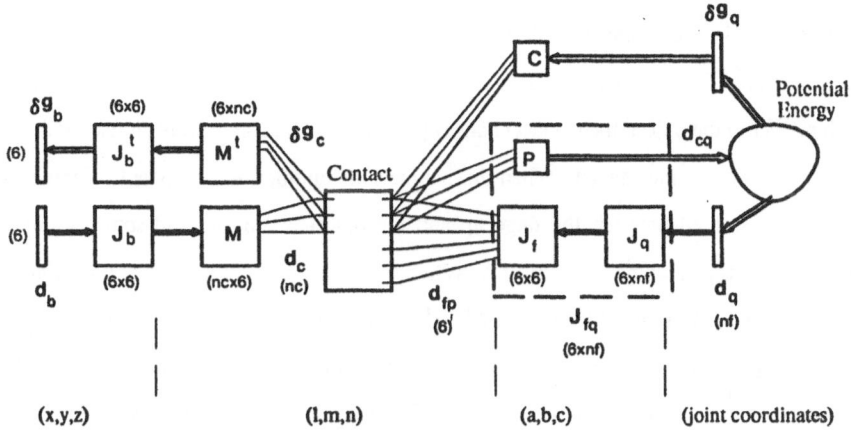

Figure 5-10: Flow chart for relationships between displacements and forces (Case 2 or 3)

5.3.5 Computing Changes in Grip Force

As mentioned in Section 5.2.3, the changes in the grip forces will be due to two effects. The first consists of the restoring forces in the finger joints produced by displacing the fingers. The second stems from relative motion between the object and the fingers which modifies the grasp geometry so that the grasp forces produce different forces and torques on the object.

The change in grip geometry can be broken into two parts. The first is due to the contact area shifting upon the object and the second is due to relative motion between the finger and the object. For the first part, recalling that the forces upon the object, g_b, are given in terms of g_{fp} by equations (5.7) and (5.8), the total change in the forces upon the object becomes

$$\Delta g_b = \Delta([Jb]^t[M]^t g_{fp}) = \Delta([Jb]^t[M]^t)g_{fp} + [Jb]^t[M]^t \Delta g_{fp}.$$

The change in the product of the jacobians above may be expanded to give terms involving $\Delta[Jb]^t$ and $\Delta[M]^t$.

$\Delta[Jb]^t$ will be zero if the contact area does not move with respect to the object when the object is displaced by d_b. This is true for point contacts and for contacts in which a very soft finger adheres to the surface. For curved finger/object contact surfaces, the contact area usually moves on the surface of the object due to rolling of the finger and $\Delta[Jb]^t$ cannot be ignored.

$\Delta[M]^t$ will be zero provided that the coupling between the finger and the object does not change. This is true for many contact geometries, although there are a few exceptions, such as a flat-ended finger touching a flat surface on the object. If the flat finger rocks slightly with respect to the object, the contact changes from a line contact to a point contact or from a planar contact to a line contact. When this happens the number of components of force and motion transmitted between the fingertip and the object is reduced and some additional elements of $[M]^t$ become zero. Such transitions, however, are not smooth and continuous and cannot be represented by a matrix $\Delta[M]^t$. In the following analysis it will be assumed that that $\Delta[M]^t$ is zero. The case of a flat-tipped finger on a flat object can be regarded as a limiting case in which the radii of curvature of the fingertip and object approach infinity.

The expression for Δg_b is now given by

$$\Delta g_b = \Delta[Jb]^t [M]^t g_{fp} + [Jb]^t[M]^t \Delta g_{fp}.$$

In Appendix A.3, a method is given for determining the elements of $\Delta[Jb]$ for a given translation and rotation of the contact area with respect to the object.

Next, it is necessary to determine the change in g_{fp}. This will be due partly to the relative motion of the finger with respect to the (l,m,n) coordinate system and partly to the restoring forces in the finger joints.

The motion of the fingertip is d_{fp}, where d_{fp} is determined by the methods of Section 5.3.3. The motion of the (l,m,n) coordinate system is given by d_{bp} and therefore the relative motion is $(d_{fp} - d_{bp})$. The resulting 6-element vector is used to determine the elements of $\Delta[Jf]^{-t}$ in the same way that the motion of the contact point on the object determines the elements of $[Jb]^t$. There may also be a contribution to Δg_{fp} due to relative motion between the (a,b,c) coordinate system and the joint space in which d_q is defined. However, this will depend on the particular finger design and is not considered in the current analysis.

The restoring forces in the finger joints are given by

$$\Delta g_q = [Kq]d_q$$

where d_q is found using the methods in Sections 5.3.1 and 5.3.3. The contribution of these restoring forces to Δg_{fp} may be computed for each of the constraint cases discussed in Section 5.3.3.

For the first case in Section 5.3.3, in which the motion of the object exactly determines the motion of the finger, the contribution of Δg_q to Δg_{fp} follows from equation (5.9).

$$\Delta g_{fp} = [P]^{-t}\Delta g_q \tag{5.11}$$

For the second case, the contribution of the restoring forces in the joints to the change in the forces at the fingertip is $[P^*]^{-t}\Delta g_q^*$. It does not matter which columns are removed from $[P]$ provided that the remaining square matrix, $[P^*]$, is non-singular and that the elements of g_q corresponding to the eliminated rows of $[P]$ are removed from g_q^*.

For the third case the problem is statically indeterminate and there are not enough equations for the number of unknowns. If no motion is possible in one of the directions of the (l,m,n) coordinate system, the change in force for that direction may reasonably be set to zero. This is equivalent to removing null rows and columns from the compliance matrix in (l,m,n) coordinates. If the remaining compliance matrix is still singular, or in other words, if there are remaining non-zero (but coupled) motions in d_{fp} then a useful technique is to add "virtual joints" to the finger to provide enough equations. The virtual joints can be chosen in directions orthogonal to the existing joints. The motion about their axes is zero and consequently, the change in the torques about their axes will also be zero.

For the particular case in which fingers with three degrees of freedom are used to hold an object, with point contact between the fingertips and the object, the relations above reduce to

$$d_q = [P]^{-1}[M][Jb]d_b$$

$$\Delta[Jb]^t = [0]$$

$$\Delta g_b = [Jb]^t[M]^t(\Delta[Jf]^{-t}g_f + [P]^{-t}g_q)$$

If, in addition, it can be assumed that $\Delta[Jf]^{-t}$ is negligible, the change in the force upon the body becomes

$$\Delta g_b = [Jb]^t[M]^t[P]^{-t}[Kq][P]^{-1}[M][Jb]d_b$$

or, letting $[P]^{-1}[M][Jb] = [J]$

$$\Delta g_b = [J]^t[Kq][J]d_b.$$

For a grip with m fingers, the net change in the grasping force becomes

$$\Delta g_b = \sum_{i=1}^{m} [J_i]^t [Kf_i][J_i] d_b$$

$$\text{or} \quad \Delta g_b = [Kb]d_b \tag{5.12}$$

Three Fingered Hand

For a hand with three fingers, each having three degrees of freedom and point contacts at the fingertips, Salisbury [78] derives an equivalent expression to (5.12). If the finger axes, (a, b, c) are chosen parallel to (x, y, z) and their origin, f, is moved to the contact point, fp, then

$$\begin{array}{ccccc} [P]^{-1} & [M] & [Jb] & = & [\ I\ |\ R^t\ ] & = & [J] \\ (3x3) & (3x6) & & & (3x6) & & \end{array} \tag{5.13}$$

where $[I]$ is a 3x3 identity matrix and $[R]$ is given in Appendix A.1. The jacobians, $[J]$, for each finger are assembled into a single grasp jacobian (*see* Figure 5-11). The 6x9 grasp jacobian is augmented by a 3x9 matrix that gives the dot products between the forces exerted by opposing fingers. These "pinch" terms are related to the magnitude of the internal forces on the object. The resulting 9x9 grasp matrix is $[G]^{-t}$. The fingertip displacements are concatenated into a single 9x1 vector d_f and the vector of resultant forces, with respect to equilibrium, on the object becomes $f_b = \Delta g_b$. The 3x3 finger stiffness matrices are also assembled into a single, block-diagonal 9x9 matrix, $[K]$.

$$[K] \quad = \quad \begin{bmatrix} Kf\ | & & \\ \hline & |\ Kf\ | & \\ \hline & & |\ Kf \end{bmatrix}$$

The relationship between displacements of the fingers and the net restoring force upon the body, f_b, may then be expressed as

$$f_b = [G]^{-t}[K]d_f$$

and the stiffness of the object computed as

$$[Kb] = [G]^{-t}[Kf][G]^{-1}$$

The relationship between the above expressions and equation (5.12) can be seen by dropping the "pinch" terms from $[G]$ and f_b, and by allowing an arbitrary number of fingers which have arbitrary orientations, $[A_i]$ with respect to the (x,y,z) system:

$$
\begin{bmatrix}
A & | & A & | & A & | & \dots \\
\hline
RA & | RA & | RA & | & \dots
\end{bmatrix}
\begin{bmatrix}
Kf & | & & & \\
\hline
& | & Kf & | & \\
\hline
& & & | & Kf & | \\
\hline
& & & & | & \dots \\
\hline
& & & & & | & Kf
\end{bmatrix}
\begin{bmatrix}
A^t & | [RA]^t \\
\hline
A^t & | [RA]^t \\
\hline
A^t & | [RA]^t \\
\hline
\dots & \dots
\end{bmatrix}
$$

Multiplying the partitioned matrices above gives

$$
[Kb] = \begin{bmatrix}
\sum_{i=1}^{m}[Kx_i] & | & \sum_{i=1}^{m}[Kx_i][R_i]^t \\
- - - & + & - - - - - \\
\sum_{i=1}^{m}[R_i][Kx_i] & | & \sum_{i=1}^{m}[R_i][Kx_i][R_i]^t
\end{bmatrix}
$$

(where $[Kx_i] = [A_i][Kf_i][A_i]^t$)

which is identical to (5.12), when $[J_i]$ are given by (5.13).

5.4 A Closer Look at Contact Conditions

Contact conditions between the gripper and the object depend on friction, adhesion, surface geometry and surface deformation under load. The contact conditions have a profound effect on the strength and stability of a grip and determine the extent of kinematic coupling between the gripper fingertips and the object.

Previous analyses [85, 78, 76, 75, 19] have used the assumption of hard surfaces and small contact areas to treat the contact areas as point contacts. This turns out to be the simplest case to handle analytically, but it becomes inaccurate when the radius of curvature of the fingertips is not small compared to the size of the object or when the fingertips deform. The effects of different assumptions concerning the fingertip geometry are shown in several examples below. In a later section, the effects of different friction models are discussed.

Models that may be used for the fingertip geometry include: point contacts, hard curved

Grasp Jacobian for Three Fingers:

$$
f_b \quad = \quad [G]^{-t} \; f_f \quad = \quad
\begin{bmatrix}
I & | & I & | & I \\
\hline
R & | & R & | & R \\
\hline
P & | & P & | & P
\end{bmatrix}
\begin{bmatrix}
f1 \\
\hline
f2 \\
\hline
f3
\end{bmatrix}
$$

or:

$$
\begin{bmatrix}
fx \\
fy \\
fz \\
-- \\
mx \\
my \\
mz \\
-- \\
p12 \\
p13 \\
p23
\end{bmatrix}
=
\begin{bmatrix}
1 & 0 & 0 & | & 1 & 0 & 0 & | & 1 & 0 & 0 \\
0 & 1 & 0 & | & 0 & 1 & 0 & | & 0 & 1 & 0 \\
0 & 0 & 1 & | & 0 & 0 & 1 & | & 0 & 0 & 1 \\
\hline
0 & -rz & ry & | & 0 & -rz & ry & | & 0 & -rz & ry \\
rz & 0 & -rx & | & rz & 0 & -rx & | & rz & 0 & -rx \\
-ry & rx & 0 & | & -ry & rx & 0 & | & -ry & rx & 0 \\
\hline
 & r12 & & | & & -r12 & & | & 0 & 0 & 0 \\
 & r13 & & | & 0 & 0 & 0 & | & & -r13 & \\
0 & 0 & 0 & | & & r23 & & | & & -r23 &
\end{bmatrix}
\begin{bmatrix}
f1x \\
f1y \\
f1z \\
-- \\
f2x \\
f2y \\
f2z \\
-- \\
f3x \\
f3y \\
f3z
\end{bmatrix}
$$

In the above:

[R] are cross-product matrices such that if $r = (rx, ry, rz)$ are vectors from the origin of the global coordinate system to each of the finger contact points, and f are three-component force vectors then $[R]f = r \times f$.

[P] are matrices formed of 3 element vectors rij which point from finger i to finger j. The products $[P]f$ produce three scalar internal forces, pij, which measure the "pinch" between fingers i and j.

Figure 5-11:
(from Salisbury [19])

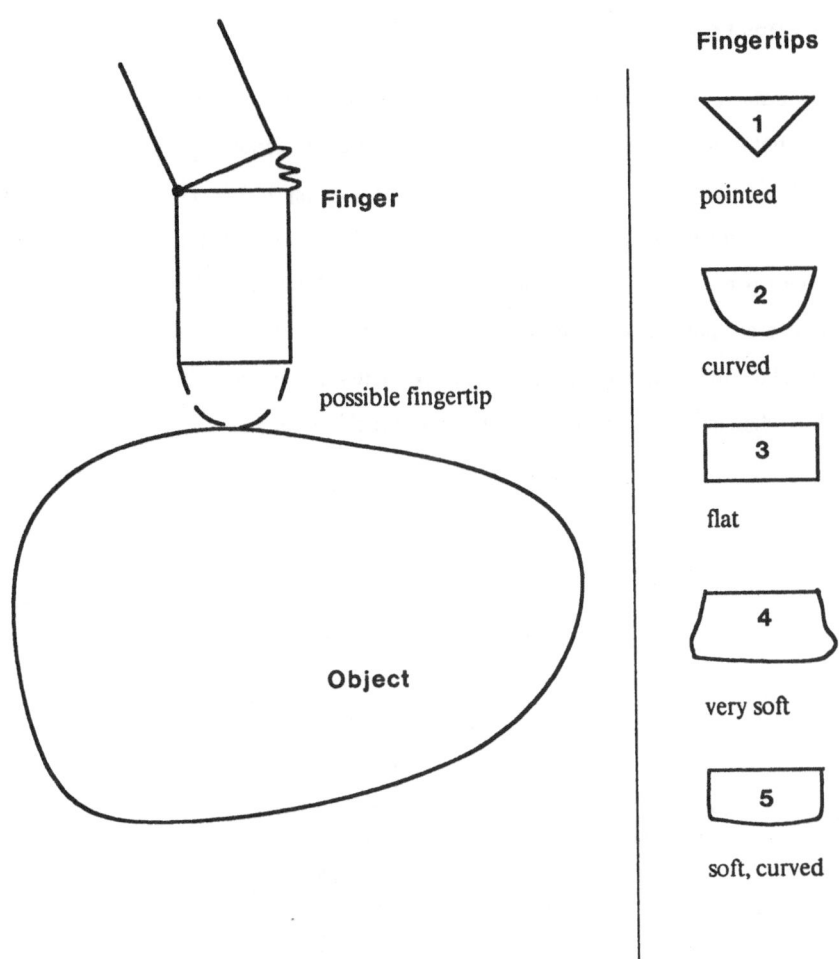

Figure 5-12: Examples of fingertip geometry

contacts, flat contacts, elastic curved contacts and very soft contacts. These models are shown schematically in Figure 5-12.

5.4.1 Point Contact

In a point contact with friction, forces are transmitted between the fingertip and the object but torques are not. Similarly, translation of the fingertip is coupled with that of the object, but rotation is not. The result is that the coupling matrix, [M], is a 3x6 matrix in which the left partition is a 3x3 identity matrix and the right partition is zero.

With point contact, there is no rolling motion and consequently no movement of the contact area upon the object or the fingertip. As the object is displaced, the fingers can only rotate about the contact points. Consequently, there is no change in the jacobian $[Jb]^t$ and only a rotational change in $[Jf]^{-t}$ as the object is displaced.

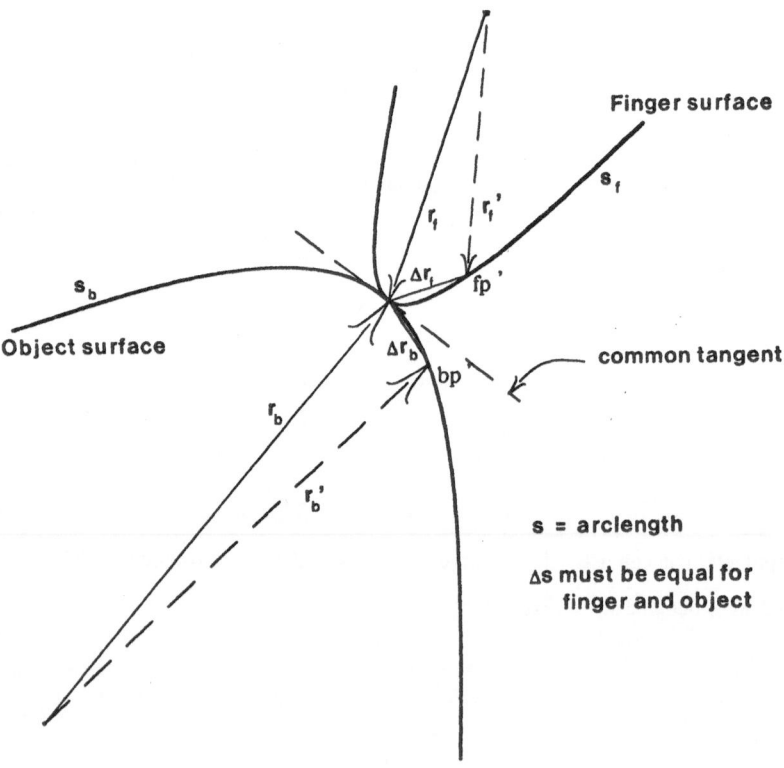

Figure 5-13: Rolling contact

5.4.2 Curved Finger Contact

A hard, curved finger is similar to a point contact in that the contact area is small so that forces may be transmitted, but torques may not. The main difference arises from the possibility of the fingertip rolling upon the surface of the object. As the finger rolls, the location of the contact point will shift. This shift produces non-zero terms in the differential jacobians, $\Delta[Jb]^t$ and $\Delta[Jf]^{-t}$ introduced in Section 5.3.5.

A general analysis of rolling becomes quite complex. As a first step, if we assume that the finger does not twist about its own axis, (perpendicular to the surface of the object) then for small displacements the problem can be approximated by a two-dimensional one involving an instantaneous plane of rolling. The plane is defined by the common perpendicular (n in Figure 5-7) and the vector of translational motion of the initial contact points, bp and fp. In the following discussion, second-order approximations are derived to express the translation and rotation of the contact points on the fingertip and the object as functions of the fingertip and object curvature.

Figure 5-13 shows the cross sections of a finger and an object in the instantaneous plane of rolling motion. The fingertip and the object profiles may be described parametrically as $r_{f(s)}$ and $r_{b(s)}$, where s is equal to the arclength along either curve. The conditions for pure rolling, without slipping or losing contact, are

1. There will be a common tangent plane at the points of contact.

2. The contact points on the fingertip and the object (fp and bp in Figure 5-7) must have the same translational velocity. For a differential motion this means that the translational components of d_{bp} and d_{fp} must be equal.

3. The arc length, δs, traversed along $r_{b(s)}$ and $r_{f(s)}$ must be equal as the fingertip rolls on the object.

The tangent at any point, s, along each curve in Figure 5-13 is given by the unit vector

$$u = \frac{dr}{ds}$$

At the contact point, the tangent is the same for both curves so that

$$\frac{dr_f}{ds} = \frac{dr_b}{ds}$$

After the fingertip rolls a small amount, the new contact point will be at the location r_b' on the body and the new tangent will have the direction

$$u' = \frac{dr_b'}{ds}$$

The contact point on the fingertip will be at the location r_f' with respect to the finger coordinate system and the direction of the tangent will be

$$u_f' = \frac{dr_f'}{ds}$$

For pure rolling it is required that $\delta s_f = \delta s_b$, where for small motions, $\delta s = \sqrt{\Delta r \cdot \Delta r}$

Thus for a small rolling motion, the contact point translates Δr_b upon the body and rotates through the angle between u_b and u_b'. At the same time, the fingertip must translate by $\Delta r_b - \Delta r_f$ (the distance between bp' and fp') and rotate through the angle between u_f' and u_b'. The translations and rotations are functions of $r_{f(s)}$, $r_{b(s)}$ and δs.

Δr and u' may be expressed as Taylor's series expansions in $r_{(s)}$ and δs (Appendix A.4). To look at the effects of curvature, terms involving the first and second derivatives of r_b and r_f are kept in the expansions.

translation of bp with respect to object:

$$\Delta r_b \approx \delta s \frac{dr_b}{ds} + \frac{(\delta s)^2}{2} \frac{d^2 r_b}{ds^2} = u \delta s + \frac{du_b}{ds} \frac{(\delta s)^2}{2} \tag{5.14}$$

translation of fp' with respect to object:

$$\Delta r_b - \Delta r_f \approx \frac{(\delta s)^2}{2} (\frac{d^2 r_b}{ds^2} - \frac{d^2 r_f}{ds^2}) \tag{5.15}$$

rotation of bp with respect to object:

$$u_b \times u_b' \approx \delta s (\frac{dr_b}{ds} \times \frac{d^2 r_b}{ds^2}) \tag{5.16}$$

rotation of fp' with respect to object:

$$u_f' \times u_b' \approx \delta s ((\frac{d^2 r_f}{ds^2} - \frac{d^2 r_b}{ds^2}) \times u) + (\delta s)^2 (\frac{d^2 r_f}{ds^2} \times \frac{d^2 r_b}{ds^2}) \tag{5.17}$$

In (5.17) and (5.14), $u = u_{b(s)} = u_{f(s)}$.

For a given object shape, the fingertip curvature determines the magnitudes of the translation and rotation of bp and the translations of fp and fp', as the fingertip rolls through the small angle given by equation (5.17).

The above equations can be simplified by dropping second order terms. Since $|u| = 1$, equation (5.17) will be dominated by a term on the order of

$$\delta s(\frac{d^2 r_f}{ds^2} - \frac{d^2 r_b}{ds^2})$$

The second term in (5.17) is at least a factor of δs smaller and, for infinitesimal motions, may be dropped. In (5.15), the translation of fp' is also smaller than the rotation of the fingertip by a factor of δs, which leads to the conclusion that for infinitesimal rolling, the fingertip may be considered to rotate about the contact point, fp. The translation of the contact point on the object, bp, contains one term on the order of δs, and a second term which may be dropped. The simplified equations are

translation of contact point with respect to object:

$$\Delta r_f \approx \Delta r_b \approx u \, \delta s \tag{5.18}$$

rotation of contact point with respect to object:

$$u_b \times u_b' \approx \delta s(\frac{d r_b}{ds} \times \frac{d^2 r_b}{ds^2}) \tag{5.19}$$

rotation of fingertip with respect to object:

$$u_f' \times u_b' \approx \delta s\left((\frac{d^2 r_f}{ds^2} - \frac{d^2 r_b}{ds^2}) \times u\right) \tag{5.20}$$

5.4.2.1 Effects of rolling motion

The meaning of the above equations becomes apparent in Figures 5-14 and 5-15, which show a finger with a curved tip of constant radius rolling on a flat surface on an object. For convenience, the coordinate systems are chosen so that (a, b), (l, m), and (x, y) all lie in the same plane. In Figure 5-14 the radius of curvature, r_c, of the fingertip is large while in Figure 5-15 it is small. In both cases $u_f = u_b = (1)i + (0)j$. Since the object is flat, the second derivative of r_b is zero.

The fingertip undergoes virtually the same motion in Figures 5-14 and 5-15, but there is a significant difference in Δr_b and Δr_f between the two cases, which stems from the difference in δs. In Figure 5-15, there is no appreciable change from r_b to r_b'. Consequently $\Delta[Jb]^t = [0]$. There is also virtually no difference between r_f and r_f', when expressed with respect to the (a, b) coordinate frame. Consequently $\Delta[Jf]^{-t}$ contains only a rotation term resulting from the rotation of the (a, b) coordinate system with respect to the contact point and the (x, y) system. In other words, as the radius of curvature becomes small, the model reduces to the case of a pointed finger rotating about its tip.

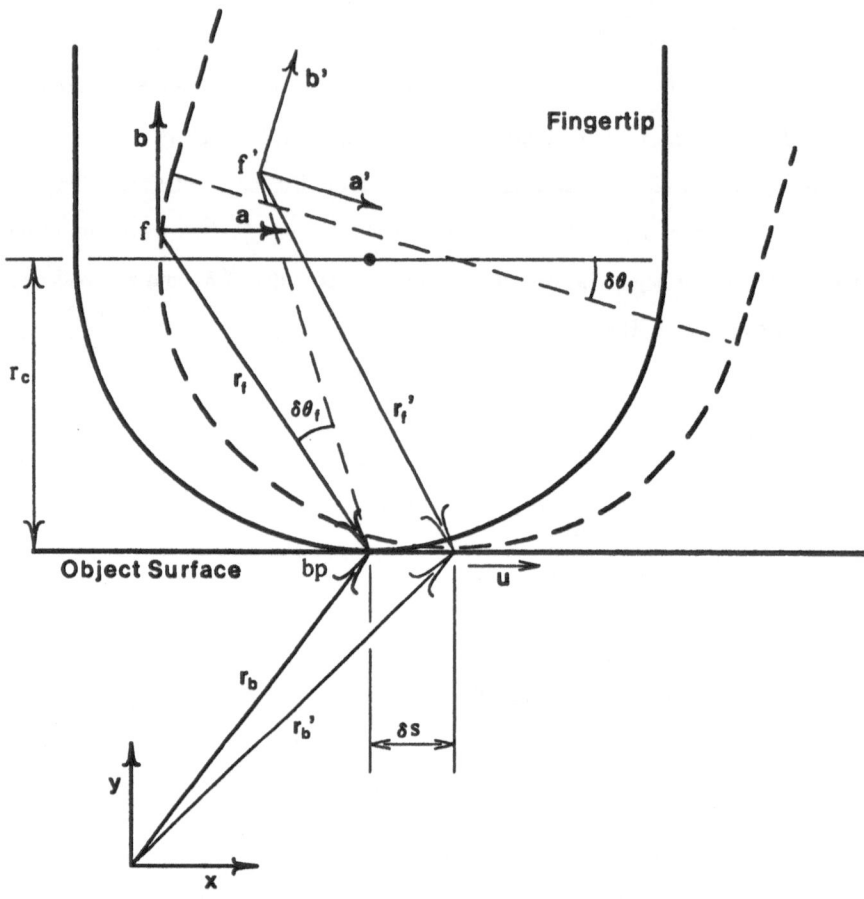

Figure 5-14: Cross section of a large-radius hemispherical fingertip on a flat object surface

In Figure 5-14, Δr_b and Δr_f are significant. Consequently, $\Delta[Jb]^t$ contains a translation term and $\Delta[Jf]^{-t}$ reflects both the rotation of the (a, b) system and the addition of Δr_f to r_f. The way in which such terms are incorporated into the elements of the differential jacobians is shown in Appendix A.3, and an example is given in Section 5.5.

A flat-tipped finger can be seen as a limiting case in which the radius of curvature becomes infinite so that Δr_b and Δr_b become infinite and produce an infinite displacement of the contact area for any rotation of the finger with respect to the object. In practice, of course, the contact point will jump to the edge of the flat fingertip, at which point the radius of curvature becomes zero rather than infinite.

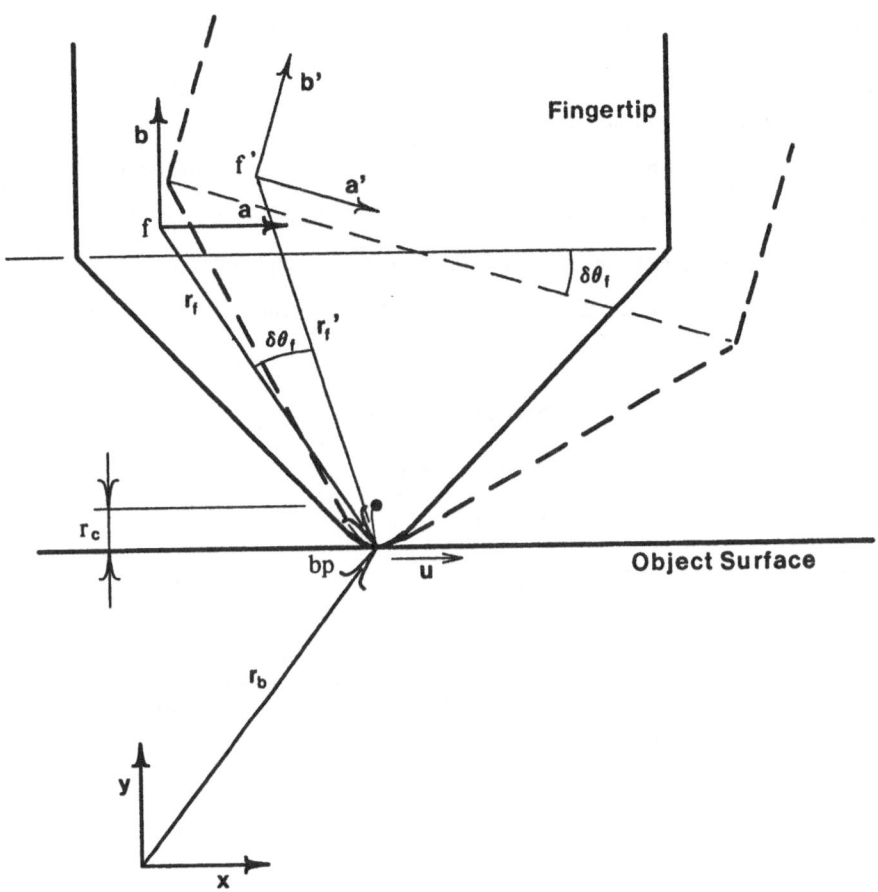

Figure 5-15: Cross section of a small-radius hemispherical fingertip on a flat object surface

5.4.3 Very Soft Finger

The fourth fingertip example shown in Figure 5-12 represents the extreme case of a compliant fingertip pressing against the object surface. In this model it is assumed that the fingertip conforms to the object surface, and adheres slightly. Such characteristics are found in many natural gripping surfaces, including the fingertips of the human hand. The coefficient of friction for such a fingertip will be high (greater than one). However, since deformation and adhesion are the primary mechanisms, it is not advisable to use the Coulomb friction law.

The soft finger model is further specialized with the assumption that no rolling occurs and that the compliant medium at the fingertip is elastic. With these assumptions the fingertip becomes a less accurate model of human fingertips. Human skin is visco-elastic and after being deformed will not generally return to its original position. Depending on the curvature of the object being held and the degree of adhesion present, the human fingertip will also roll slightly upon the object, exhibiting a rolling resistance of the kind discussed in Section 5.4.4. Nonetheless, the elastic soft-finger model is useful to demonstrate a limiting case in which there is kinematic coupling between the fingertip and the object in all six degrees of freedom.

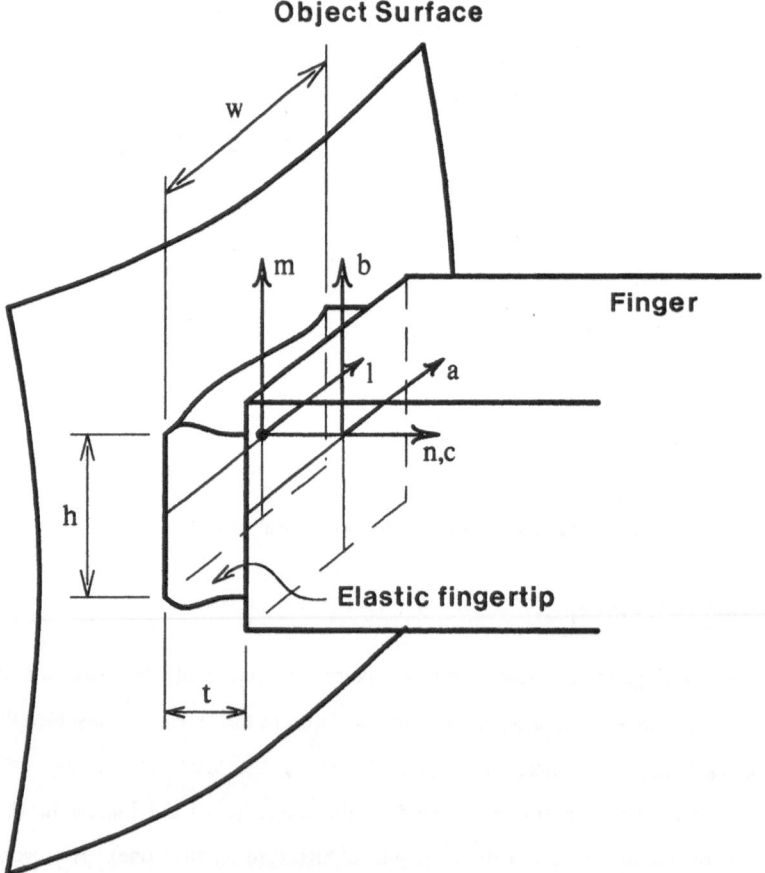

Figure 5-16: Elastic fingertip in contact with object surface (perspective)

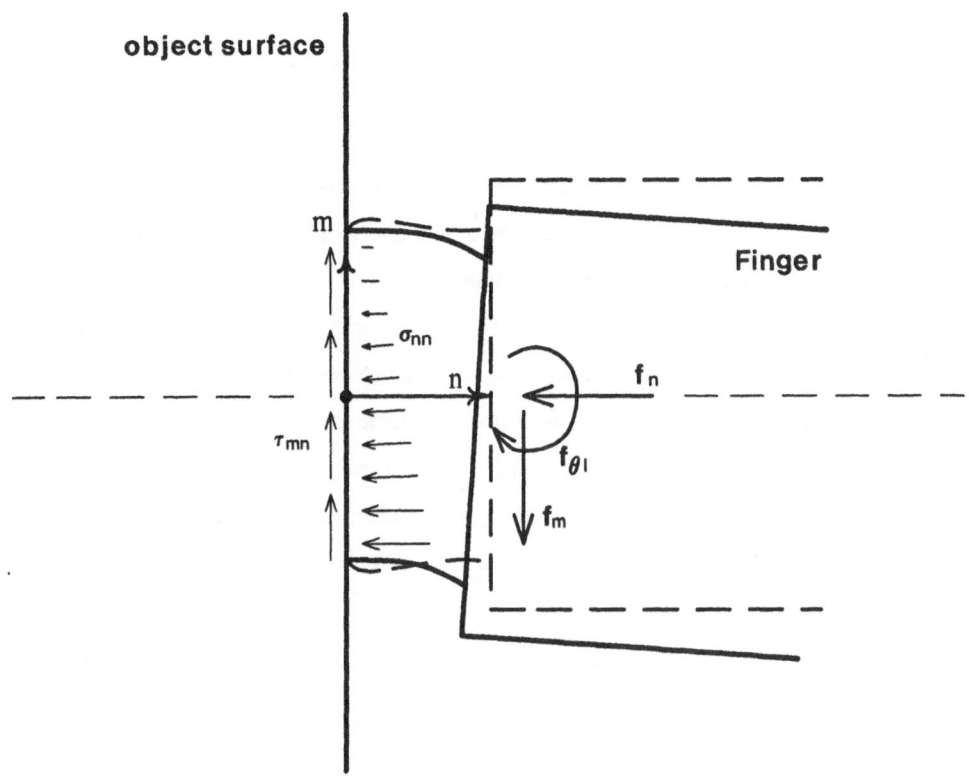

Figure 5-17: Elastic fingertip in contact with object surface (section)

Figure 5-16 shows the fingertip in contact with the surface of an object. The fingertip material is assumed to be much softer than that of either the object or the finger substrate, which are treated as rigid bodies. For convenience, the finger (a, b, c) coordinate system has been moved to the interface between the fingertip material and the finger substrate. As explained in Section 5.3.1, the forces, displacements and stiffness characteristics of the finger can easily be transformed from another coordinate system to the one chosen in Figure 5-16. The (l, m, n) coordinate system remains, as usual, at the contact area between the fingertip and the object, with the n axis normal to the object surface.

The grasping forces at the object surface, g_{bp}, can be expressed as integrals of the stresses over the contact area:

$$g_l = \int_A \tau_{ln} \, dA \qquad g_m = \int_A \tau_{mn} \, dA \qquad g_n = \int_A \sigma_{nn} \, dA$$

$$g_{\theta l} = \int_A m \, \sigma_{nn} \, dA \qquad g_{\theta m} = \int_A l \, \sigma_{nn} \, dA \qquad g_{\theta n} = \int_A (\sqrt{l^2 + m^2}) \, \tau_{lm} \, dA$$

The elastic contact represents a compliant coupling in which small motions of the finger with respect to the object are possible in any direction. Such relative motions produce changes in the above forces and a model of the system permits the deflection/force relationships to be expressed as the stiffness of the contact.

The fingertip can be treated as a short elastic member clamped between two rigid boundaries. To obtain the exact stress field for such a problem is a formidable task — even if the assumption is made that the material is perfectly elastic and isotropic. Numerical results could be obtained using a finite element analysis, but the analysis would be time consuming and would have to be re-computed for different cross sections and materials. The problem can be simplified by observing that the stresses at any given location within the material are of little interest, provided that

- estimates of the integral quantities can be computed at the object surface

- the combined stress field nowhere exceeds the strength of the material

- the normal stress, σ_{nn}, never becomes sufficiently tensile to cause the fingertip material to separate from the object surface.

The last requirement can be satisfied by assuming a large grasping force normal to the object surface and/or some adhesion between the fingertip and the object. If one edge of the fingertip *does* start to separate from the object when the finger rotates slightly, the finger is starting to roll.

Since an exact elastic solution is impractical (and would in any event be an approximation to the visco-elastic behavior of compliant polymers and skin-like materials) an approximate elastic solution is used to estimate the force/deflection relationship for the fingertip. The behavior of the fingertip in shear, torsion, compression and bending is discussed below, and the separate solutions are superposed to produce a 6x6 stiffness matrix for the contact.

Bending stiffness and resistance to rolling

The bending model for the elastic fingertip is similar to that used in classical beam theory. A rotation about the a axis by the finger produces a rotation in the material of $\delta\theta/t$ per unit thickness. The bending strain and stress at a distance m above the centerline are

$$\varepsilon_{nn} = \frac{m\delta\theta}{t} \quad \text{and} \quad \sigma_{nn} = E\varepsilon_{nn}$$

where E is the modulus of elasticity. As in beam theory, it is assumed that plane sections remain plane and $\gamma_{ln} = \gamma_{mn} = 0$. It is also assumed that since the stresses τ_{lm}, σ_{mm} and σ_{ll} are zero at the surfaces of the material, they are approximately zero throughout. This assumption is somewhat less supportable than in beam theory since the elastic element cannot be considered slender. However, it is not actually necessary that τ_{lm}, σ_{mm} and σ_{ll} be zero everywhere but only that their resultant does not significantly affect the estimated bending rigidity of the element. The bending rigidity may then be found by equating the energy stored in rotating the finger with the energy stored in deforming the material.

$$\tfrac{1}{2}k_{\theta l}(\delta\theta_l)^2 = \tfrac{1}{2}\int_V \sigma_{nn}\varepsilon_{nn}\,dV \quad = \quad \frac{EI_{mm}(\delta\theta_l)^2}{2t}$$

$$\text{or} \quad k_{\theta l} = \frac{EI_{mm}}{t} \tag{5.21}$$

where I_{mm} is the moment of inertia of the cross section about the m axis and V is the volume of the material.

The bending stiffness for rotations about the m axis is similarly found as

$$k_{\theta m} = \frac{EI_{ll}}{t} \tag{5.22}$$

As mentioned earlier, the maximum bending moment that the contact can sustain is limited by the adhesion between the fingertip and the object surface. The limitation is easily demonstrated for the example of a square contact area of length w on each side. Referring to Figure 5-17, a normal force of magnitude f_n produces a uniform contribution to the normal stress of

$$\sigma_{nn} = -\frac{f_n}{w^2} \quad \text{(compression)}$$

A bending moment of magnitude $f_{\theta l}$ produces a contribution to the normal stress that is maximum at the edges of the contact.

$$\sigma_{nn} = \pm \frac{w f_{\theta l}}{2I} = \pm \frac{6 f_{\theta l}}{w^3} \quad \text{(bending)}$$

The combined normal stress will become tensile at one edge when

$$f_{\theta l} > \frac{w f_n}{6} . \tag{5.23}$$

Thus, unless the adhesion between the fingertip and the object is able to resist tensile loads, the finger will start to roll whenever the bending moment is more than one sixth the normal load times the length of the side. For small contact areas the fingertip is likely to start rolling unless considerable adhesion is present.

Shear stiffness and resistance to slipping

For a beam with an end load, the variation in the moment over the length of the beam is balanced by a distribution in shearing stress over the cross section of the beam [114]. For the elastic fingertip, however, it is assumed that the variation in the moment produced by a shear force in the (a,b,c) system is negligible compared to the effect of rotating the finger. Consequently the bending moment is approximately constant over t and the shear stress is assumed to be uniform over the cross section. The shear stiffness is found by equating the energy required to displace the finger in shear with the energy stored internally in the material.

$$\tfrac{1}{2} k_{\theta m} (\delta_m)^2 = \tfrac{1}{2} \int_V \tau_{mn} \varepsilon_{mn} \, dV = G A \frac{(\delta \theta_l)^2}{2t} \tag{5.24}$$

$$\text{or} \quad k_m = \frac{G A}{t}$$

In the above, G is the shear modulus of the material and A is the cross section area, wh.

The maximum shearing force that the contact can sustain is limited by the shear strength of the fingertip/object interface, which depends on the bonding strength between the fingertip and object materials and on the area of intimate contact between them. The area of intimate contact is generally much smaller than the overall contact area, A, and depends not only on the current normal force, f_n, but on such factors as how clean the surfaces are, how rough they are, and how long they have been held together. In general, the shear strength of the contact will be some fraction, β, of the shear strength of the fingertip material. The fraction will be a function of (but not directly proportional to) the normal force, and slipping will occur when τ_{mn} or τ_{ln} exceeds that fraction.

$$\tau_{slip} = \beta_{(f_n, \ldots)} \tau_{yield}$$

Compressive stiffness

Displacement of the fingertip toward the object results in a uniform compressive strain, ε_{nn}, across the cross section. The compressive stiffness is found in the same way as the shear stiffness, with G replaced by E.

$$k_n = \frac{E A}{t} \tag{5.25}$$

Torsional stiffness and resistance to slipping

The torsional rigidity per unit length of a cylindrical member can easily be found as

$$k_{\theta n} = \frac{\pi G r_o^4}{2} = G I_p \tag{5.26}$$

where r_o is the radius of the cylinder and I_p is the polar moment of inertia [114]. For non-circular cross sections the expression becomes more complicated due to warping of the cross sections, although for the present case the warping may be negligible since t is small and since the material is constrained by a rigid boundary at each end. For a bar of elliptical cross-section the torsional rigidity per unit length has been determined as

$$k_{\theta n} = \frac{G A^4}{4\pi^2 I_p} \tag{5.27}$$

and it has been found that this formula holds approximately true for other compact cross sections [115].

For a round bar, the shear stress in torsion is

$$\tau_{lm} = \frac{r f_{\theta n}}{I_p} \tag{5.28}$$

Thus, if the fingertip were a cylinder ending in a circular contact area, slipping would begin at the periphery when

$$f_{\theta n} = \frac{\tau_{slip} I_p}{r_0} \tag{5.29}$$

(where τ_{slip} is given above for shear loading.) Once slipping has occurred at the periphery, the fingertip will not return to exactly the same orientation when the torque is removed. As the torque is increased, the region of slipping will spread from the periphery toward the center. The phenomenon resembles the yielding of an elastic/perfectly plastic bar in torsion. At any stage, the moment balance is given by

$$f_{\theta n} = \int_0^{r_{slip}} 2\pi \, \tau_{lm} r^2 \, dr + \int_{r_{slip}}^{r_0} 2\pi \, \tau_{slip} r^2 \, dr \tag{5.30}$$

The above equation can be integrated and condensed by expressing τ_{lm} and r_{slip} in terms of r and the angle of rotation of the finger, $d_{\theta n}$.

$$\tau_{lm} = \frac{2\pi G d_{\theta n} r}{l}$$

$$r_{slip} = \frac{l \, \tau_{slip}}{G d_{\theta n}}$$

The result for the torque is

$$f_{\theta n} = \tfrac{2}{3} \pi \, \tau_{slip} (r_0^3 - \tfrac{1}{4} r_{slip}^3) \tag{5.31}$$

Thus, the torque required for complete slipping is $\tfrac{4}{3}$ the torque required to initiate slipping at the periphery, although this would theoretically only be reached for an infinite rotation, $d_{\theta n}$ or $d_{\theta c}$, of the fingertip. For a square or rectangular contact area the qualitative behavior is the same, with slipping initiating at the periphery and spreading inwards. However, the expression for $f_{\theta n}$ becomes more complex due to the more involved expression for τ_{lm}.

Fingertip Stiffness Matrix

The above stiffness terms form the elements of a 6x6 diagonal matrix [Kc] where

$$Kc_{11} = Kc_{22} = \frac{GA}{t} \tag{5.32}$$

$$Kc_{33} = \frac{EA}{t} \tag{5.33}$$

$$Kc_{44} = \frac{EI_{mm}}{t} \tag{5.34}$$

$$Kc_{55} = \frac{EI_{ll}}{t} \tag{5.35}$$

$$\frac{GA^4}{4\pi^2 I_p t} < Kc_{66} < \frac{GI_p}{t} \tag{5.36}$$

If t were larger than w and h, then shear loads would produce bending moments that varied along t, and bending loads would produce shear deflections, as in classical beam theory. The result would be off-diagonal terms in [Kc].

5.4.3.1 Effects of deforming fingertips

The comparative importance of the above quantities can be determined for a fingertip of given proportions. The table below shows the results for two fingertips. For the first, $w = h = 1.0$ cm and $t = 0.5$ cm. In the second $w = h = 2.0$ cm and $t = 0.5$ cm. The modulus of elasticity, E, is assumed to be 250 N/cm^2 and Poisson's ratio is taken as 1/2, so that $G = E/3$. These are typical values for soft rubber. A force of 4.0 N (a little less than one lbf) is used to produce deflections for comparison.

For the smaller area, the rotational stiffness terms are much lower that the translational terms and the fingertip is clearly less constrained with respect to rotations than translations. However, the bending and torsional stiffnesses increase as the square of the contact area, while the shear and compressive stiffness increase linearly with the contact area. Thus, for the larger contact patch, the rotational and translational stiffnesses become comparable. If w and h were doubled again, bending and torsional deflections would become small in comparison to shear deflections. This result matches what one would expect intuitively.

If the grasping force is varied proportionately with the contact area, then, as the contact area becomes small, the fingertip begins to behave like a point contact in which significant

Table 5-1: Soft fingertip deflections for 4.0 N load and 1 cm^2 or 4 cm^2 contact area

Fingertip material properties: $E = 250 \text{ N/cm}^2$, $\nu = 0.5$, $G = 83.3 \text{ N/cm}^2$

	$w = h = 1.0$ cm, $t = 0.5$ cm	$w = h = 2.0$ cm, $t = 0.5$ cm
Kc_{11}, Kc_{22}	167 N/cm	667 N/cm
deflection for 4.0 N shear force	0.024 cm	0.006 cm
Kc_{33}	500 N/cm	2000 N/cm
deflection for 4.0 N compressive force	0.008 cm	0.002 cm
Kc_{44}, Kc_{55}	42 Ncm	672 Ncm
rotational deflection for 4.0 N at 1.0 cm lever arm	0.096 radian	0.006 radian
Kc_{66}	27 Ncm	432 Ncm
torsional deflection for 4.0 Ncm torque	0.15 radian	0.009 radian

rotations are possible but translations are not. As the contact area becomes large, rotations are negligible compared to shear deflections. If the grasping force is held constant for different contact areas then the contact becomes much less compliant as the area increases, and rotational deflections become negligible faster than translational deflections.

For the forces given in the table above, unless some adhesion exists between the fingertip and the object, the bending moment will cause the fingertip to roll for both the 1 cm^2 or the 4 cm^2 area. From equation (5.23), the largest bending moment that the contact could sustain without tensile stresses occurring is 0.67 Ncm for the 1 cm^2 case and 1.33 Ncm for the 4 cm^2 case. In torsion, depending on the shear strength of the interface, the contact will probably slip for the 1 cm^2 area but might not for the 4 cm^2 area. From equations (5.29) and (5.31), if the shear strength is roughly equal to 4.0 N/cm^2 in the first case (corresponding to a coefficient of friction of 1.0) and 1.5 N/cm^2 in the second, (corresponding to a coefficient of friction of 1.5)[6] the maximum torques that can be exerted are 1.7 Ncm and 5.3 Ncm respectively. This supports the idea that a soft finger with a small contact area can exert torques about an axis normal to the contact surface more readily than it can exert torques in the plane of the surface. For a soft, curved fingertip, as discussed below, the difference is more pronounced.

Once the fingertip stiffness matrix has been computed, the net compliance matrix may be formed by adding the compliances for the finger and the fingertip.

$$[Cf] = [Jfq][Kq]^{-1}[Jfq]^t + [Kc]^{-1}.$$

This matrix is invertible and therefore, the restoring forces at the contact become

$$\Delta g_{fp} = [Cf]^{-1}d_{bp}$$

Using equations (5.5) and (5.6), the changes in the forces at the finger joints are $\Delta g_q = [Jfq]^t \Delta g_{fp}$, and the finger motions are $d_q = [Kq]^{-1}\Delta g_q$.

5.4.4 Soft, Curved Fingertip

The hard curved fingertip and the very soft fingertip represent extremes between which real, deformable fingertips may be expected to lie. Human fingers and rounded robot fingers with rubber surfaces exhibit both rolling and substantial deformation. The analysis of such fingertips becomes quite involved, combining the rolling calculations of Section 5.4.2 with the deformation calculations of Section 5.4.3. A complete model is not attempted in the discussion below, but the properties of soft, rolling fingers are discussed and it is seen that they are bracketed by the models developed in the last two sections.

A number of insights can be gained by considering the analyses applied to rolling rubber tires and metal cylinders or spheres. For a hard, elastic sphere rolling on an elastic surface, the pressure distribution is described by the Hertzian contact model of solid mechanics, which predicts a hemispherical pressure distribution [115]. For the much larger

---

[6]According to the Coulomb theory, the coefficient of friction is independent of the contact area, but for compliant materials it is generally not independent.

deformations that occur when a soft, curved finger presses against an object the distribution is expected to be only qualitatively similar. The pressure will be maximum at the center of the contact, diminishing smoothly to zero at the periphery. For progressively softer fingertips, the pressure distribution becomes more uniform, especially toward the center of the contact area. In the limiting case, the pressure is essentially uniform throughout, as assumed in the very soft finger model described in Section 5.4.3. The pressure distributions are compared for elastic, soft, and very soft fingertips in Figure 5-18.

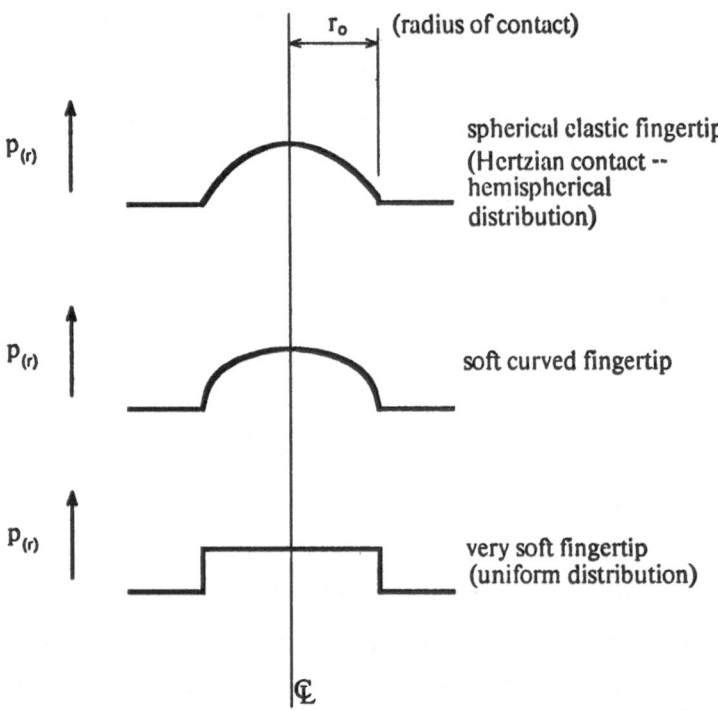

Figure 5-18: Pressure distributions for elastic, soft, and very soft fingertips

For a perfectly elastic curved finger, it is impossible to transmit moments in the plane of the contact since the finger rolls easily upon the object. Thus, in the absence of rolling resistance, the soft curved finger would behave in the same manner as the hard curved finger discussed earlier, the only difference being that r_f would vary due to flattening of the fingertip under load. If the degree of flattening could be predicted as a function of fingertip loading, then the methods discussed in section 5.4.2 could be used to predict the motion of the finger and the contact point. Elastic flattening formulas have been developed for cylinders and spheres, but these are unlikely to give accurate results for a soft fingertip.

In practice, there is generally a resistance to rolling. At low speeds, the rolling resistance is due largely to hysteresis losses and microslip at the contact area. Rolling resistance is an important subject in the literature on wheels and tires and is discussed at length in [116, 117, 118]. For an elastic sphere or cylinder rolling upon a plane surface, the deformation of the material results in a hysteresis loss which can be used to derive a "coefficient of rolling resistance" [116]. Microslip results from the elastic strain of the fingertip material as it is pressed against the surface. If the fingertip is loaded with a normal load, f_n, against the object surface, the material ahead of the centerline of the contact will spread forwards slightly and the material behind the centerline will spread backwards slightly. The spreading produces regions of microslip toward the front and rear of the contact area. In the absence of tangential forces, the strains, and the shear tractions that result from them, must cancel each other. When a tangential force is present, there will be a region of sticking toward one side of the contact area, and microslip elsewhere. The microslip results in rolling losses and "creep." The end result is that soft curved fingertips do not rotate as freely with respect to the object as pointed or hard curved fingertips do.

The static resistance to slipping of a soft, curved fingertip will be similar to that of the very soft finger in Section 5.4.3, except that since the pressure distribution is not uniform over the contact area, the value of the stress at which slipping occurs also varies over the contact. As in Section 5.4.3, the interface shear strength τ_{slip}, may be expressed as a fraction of the material shear strength, where the fraction, β, is a function of factors including the normal pressure and the surface roughness. Since the pressure is least at the edges of the contact, slipping may be expected to initiate there.

For loads in the plane of the contact, the shear stress may be uniform inside the region where there is no sliding, but will have an upper limit of τ_{slip} outside the region.

For a moment about the axis normal to the contact, the shear stress inside the sticking region will have the same distribution as for the very soft finger, the magnitude at any point being proportional to the distance from the center of the contact as in equation (5.28). In the

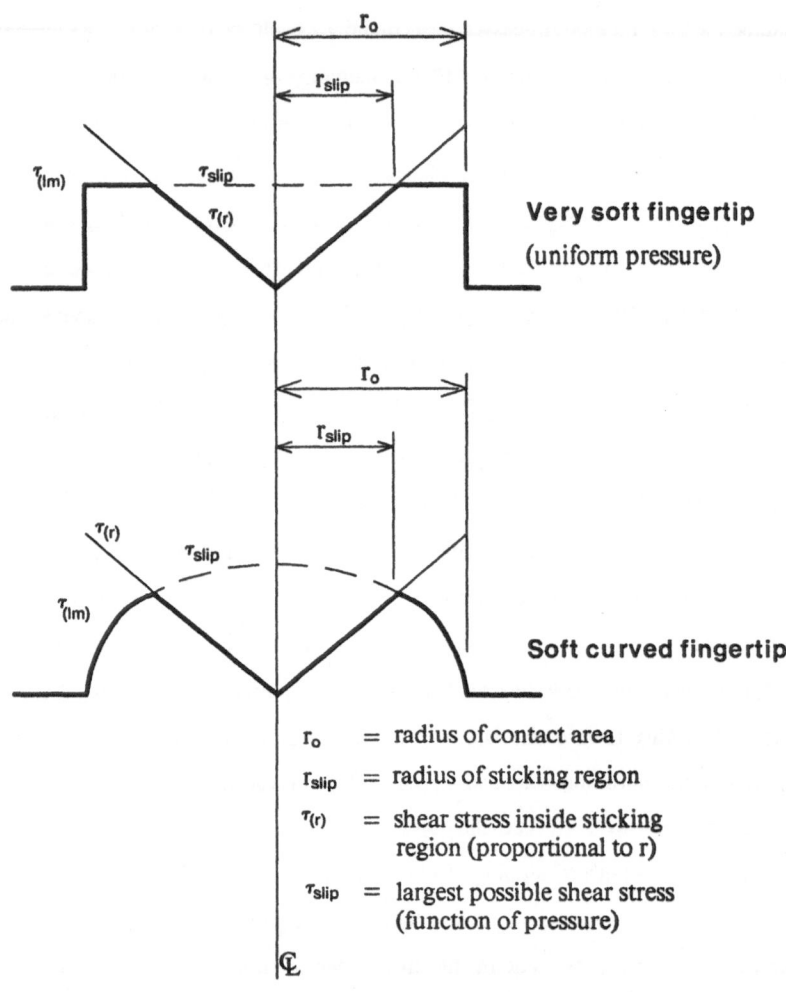

Figure 5-19: Maximum shear stress for moment about finger axis

slipping region, the shear stress will again be equal to the upper limit of τ_{slip}. A cross section of shear stress distribution is shown in the lower part of Figure 5-19. The distribution for the very soft fingertip of Section 5.4.3 is shown in the upper part for comparison. The maximum torque about the axis of the finger is equal to the polar moment of the shear stress shown in Figure 5-19

$$f_{\theta n} = \int_0^{r_{slip}} 2\pi \, \tau_{lm} \, r^2 \mathrm{d}r + \int_{r_{slip}}^{r_o} 2\pi \, \tau_{slip} \, r^2 \mathrm{d}r \tag{5.37}$$

Thus, unlike the hard curved finger or the pointed finger, the soft curved finger is able to exert small torques about its own axis.

5.5 Examples

In this section, results from the last three sections are used in three examples that also illustrate some of the differences between pointed, curved and soft fingers. Figure 5-20 shows three rectangles, each held by two fingers. In the first case the fingers are pointed, in the second case they have finite radii of curvature and in the third case they have very soft tips that adhere to the surface of the rectangle. In all three cases, the fingers are assumed to have three degrees of freedom, being restricted to motions within the plane of the paper. For simplicity, it is assumed that the finger joints correspond to translations, a and b, and a rotation, θc, in the (a, b, c) frames.

The sizes and orientations of the rectangular object and fingers, and the finger stiffnesses, $[Kf]$, are identical in each case. However, the different contact conditions produce substantially different results for the mobility, stiffness, strength and stability of the grasp.

In each case the change in the resultant grasp force on the object, Δg, is calculated for small displacements of the object. The grip stiffness is computed and the maximum force and torque that the grip can resist without slipping is calculated.

5.5.1 Pointed Fingers

The transformation matrices, $[Jb]$, $[M]$, $[Jf]$, and $[Jq]$, are given in Appendix A.5 for the left or first finger. As the object is moved an arbitrary amount, d_b, the motions at the contact points on the object are given by $d_{bp} = [Jb]d_b$. Premultiplying by $[M]$ gives the vectors d_c, which contain just the first three elements of d_{bp} since, for point contact conditions, only the translations are transmitted.

The fingers have three degrees of freedom and consequently $d_q = [dq_a, dq_b, dq_c]$. A motion, d_q, produces a motion d_{fp} at the fingertip, given by equation (5.3). The elements of d_{fp} and d_c are compared below for the left and right fingers. The (l, m, n) coordinate systems are shown in Figure 5-20.

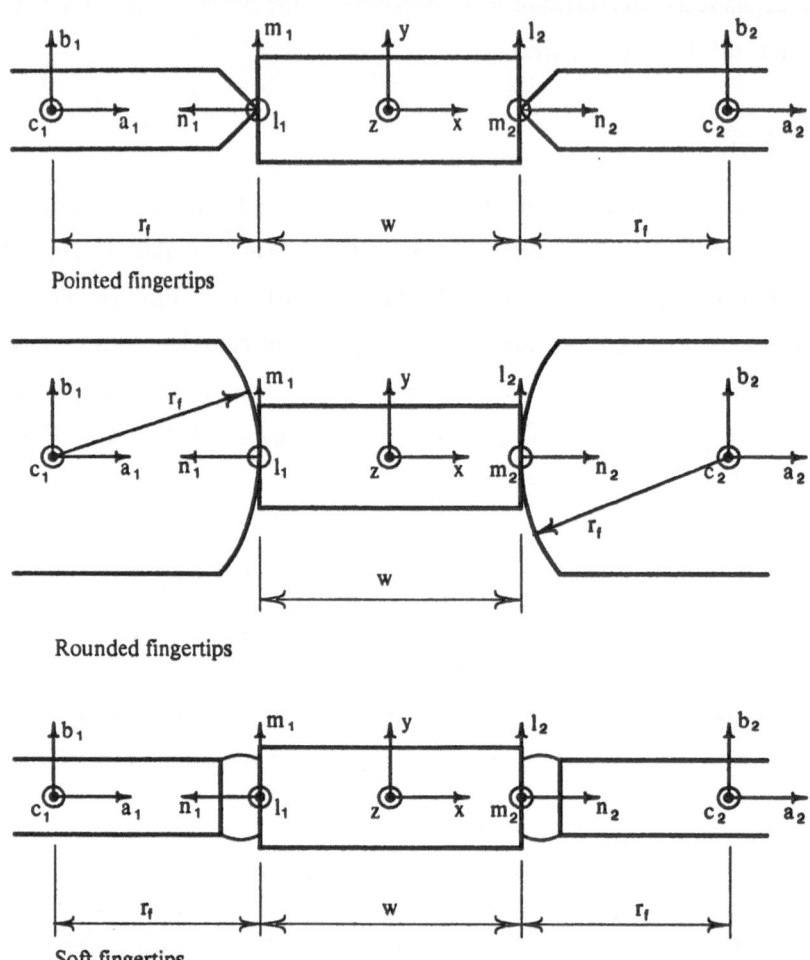

Pointed fingertips

Rounded fingertips

Soft fingertips

Figure 5-20: Holding a rectangle between two fingers — 3 examples

	d_{c1}	d_{fp1}	d_{fp2}	d_{c2}
dl	$dz + \frac{w}{2}d\theta y$	0	$dq_b - rf\,dq_c$	$dy + \frac{w}{2}d\theta z$
dm	$dy - \frac{w}{2}d\theta z$	$dq_b + rf\,dq_c$	0	$dz - \frac{w}{2}d\theta y$
dn	$-dx$	$-dq_a$	dq_a	dx
$d\theta l$		dq_c	0	
$d\theta m$		0	dq_c	
$d\theta n$		0	0	

Table 5-2: Motions of left and right finger and object contact areas (pointed fingertips)

Matching d_{c1} with d_{fp1} and d_{c2} with d_{fp2} reveals that $dz + \frac{w}{2}d\theta y = 0$ and $dz - \frac{w}{2}d\theta y = 0$ or, $dz = d\theta y = 0$. In other words, the object is constrained by the fingers to move within the plane, except for rotations about the x axis. In the following discussion it will be assumed that the object is displaced in the x and y directions and rotated about the z axis. Thus, dl_1 and dm_2 will be zero and the only motions transmitted to the fingers will be dm_1, dn_1, dl_2 and dn_2.

The procedure for calculating the finger motions, the changes in the finger forces and the change in force on the body is given below for the first finger. The contribution from the second finger follows from symmetry.

5.5.2 Procedure for Left Finger

The first step is to determine the motions of the first finger given dm_1 and dn_1. A motion in the dm direction can be accommodated either by a movement of the finger in the b direction or by a rotation about c. In practice, both will occur and the contribution from each will be balanced to minimize the potential energy of the finger. The two rows of [Jfq] that relate finger motions dq_b and dq_c to dm and dn are extracted to form the 3x2 matrix [P]

Following the method in Section 5.3.4.2, a Lagrange multiplier matrix, $[L]$ is assembled from $[Kq]$ and $[P]$. The matrices and the matrix equations are shown in Appendix A.5. Inverting $[L]$ produces the finger motions, d_q. Multiplying the finger joint motions by $[Kq]$ determines the changes in the joint forces.

The changes in the forces at the fingertip, δg_{fp}, depend both on the restoring forces δg_q and the change in geometry, $\Delta[Jf]^{-t}$, due to the motion of the finger with respect to the object. The motion of the (l, m, n) coordinate system is given by d_{bp} and the motion of the fingertip is given by $[Jfq]d_q$. The translations of each are the same, but the finger rotates relative to the object by the angle

$$\delta\theta = \delta\theta l - \delta\theta c.$$

which appears as a rotation term in $\Delta[Jf]^{-t}$ in Appendix A.5.

For a grasping force of f in the a direction and for a motion $(dx, dy, d\theta z)$ applied to the body, the change in the force applied by the first finger to the object is shown in Table 5-3.

$$\delta g_x \quad = -k_a\,dx$$

$$\delta g_y \quad = (\alpha f - \beta)\,dy - (f + \tfrac{w}{2}(\alpha f - \beta))\,d\theta z$$

$$\delta g_z \quad = 0$$

$$\delta g_{\theta x} \quad = 0$$

$$\delta g_{\theta y} \quad = 0$$

$$\delta g_{\theta z} \quad = -\tfrac{w}{2}(\alpha f - \beta)\,dy + \tfrac{w}{2}(f + \tfrac{w}{2}(\alpha f - \beta))\,d\theta z$$

$$\text{where} \quad \alpha = \frac{k_b r_f}{k_b r_f^2 + k_c} \quad \text{and} \quad \beta = \frac{k_b k_c}{k_b r_f^2 + k_c}$$

Table 5-3: Contribution from left finger to δg_b (pointed fingertip)

When the second finger is added, the expressions for the change in the force on the object become simpler due to combinations and cancellations of symmetrical terms. The contributions to δg_y from each finger cancel for rotations, $d\theta z$, and add for translations, dy. Similarly, the contributions to $\delta\theta z$ from each finger cancel for motions, dy, and add for rotations $d\theta z$. The final result is given in Table 5-4.

$$\delta g_x \quad = -2\,k_a\,dx$$

$$\delta g_y \quad = 2(\alpha f - \beta)\,dy$$

$$\delta g_{\theta z} \quad = w(f + \tfrac{w}{2}(\alpha f - \beta))\,d\theta z$$

Table 5-4: Change in g_b due to motions dx, dy, and $d\theta z$ (pointed fingertips)

5.5.2.1 Discussion

Whenever any of the forces in Table 5-4 becomes positive, the grasp will be unstable for infinitesimal displacements in the corresponding direction. Thus, if k_c is small, $(k_c < f r_f)$, the change in the grasp force for a motion in the y direction will be positive, tending to continue the motion. This result matches one's intuition that a rectangle squeezed between two fingers will be unstable if the fingers pivot freely, without springs, about axes c_1 and c_2.

Similarly if $\tfrac{w}{2}\beta < (f + \tfrac{w}{2}\alpha f)$, the rectangle will be unstable with respect to rotations about the z axis. This result is less intuitively clear but it becomes apparent if k_c is very large, in which case the fingers do not rotate about their c axes. For this case, $\alpha \to 0$ and $\beta \to k_b$. If the object is rotated by $d\theta_z$ the change in the torque upon the body is

$$w(f - \tfrac{w}{2}k_b)\,d\theta z$$

This is exactly the result obtained earlier in Section 1 for the rotation of a rectangle squeezed between two fingers, where $k_b = k_l$ and $\tfrac{w}{2} = r$.

5.5.3 Curved Fingertips

Most of the results from the last example also apply for fingers with curved surfaces. The difference is that the contact point is no longer fixed with respect to the object and consequently $\Delta[Jf]^{-t}$ is slightly different from above and $\Delta[Jb]^{t}$ is no longer zero. The new matrices are given in Appendix A.5.

As with the pointed finger example, results are derived for the first or left finger. In the current example, the algebra has been simplified by assuming that the (a,b,c) finger

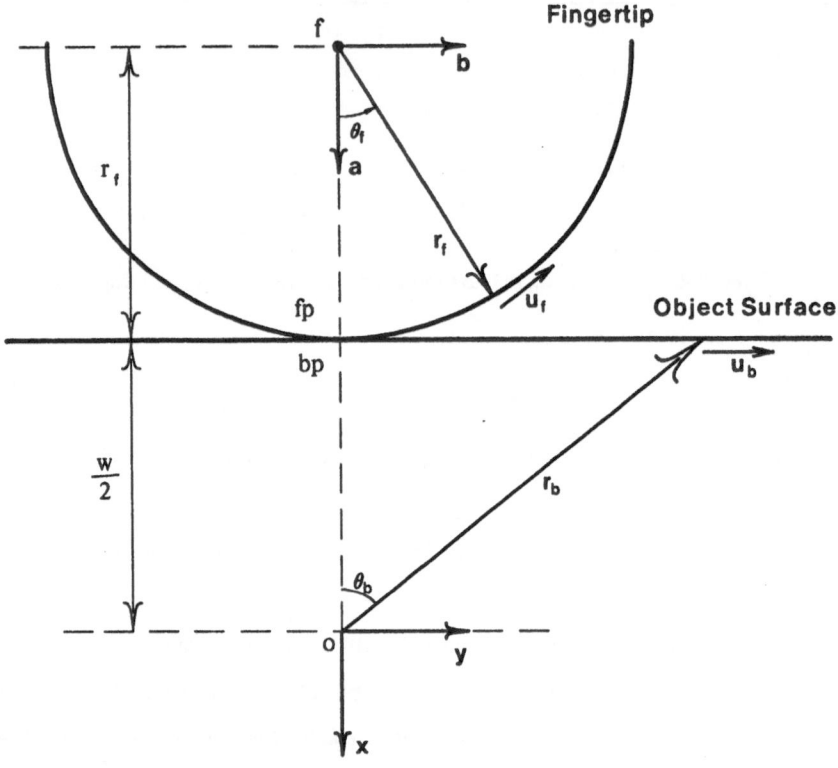

Figure 5-21: Curved fingertip

coordinate systems are also the centers of curvature of the finger tips. The rolling condition is therefore as shown in figures 5-21 and 5-22, before and after the finger has rotated a small amount, $\delta\theta$, with respect to (l,m,n) coordinate system on the object. As the figures show, $\delta s = r_f \delta\theta$.

Since the center of curvature of the finger is also the origin of the (a,b,c) system, the translation of the contact point *exactly cancels* the product $\delta\theta \times r_f$. The contribution from the left finger to Δg_b is shown in Table 5-5.

When the results from the second finger are added, the changes in the force upon the object are as shown in Table 5-6.

115

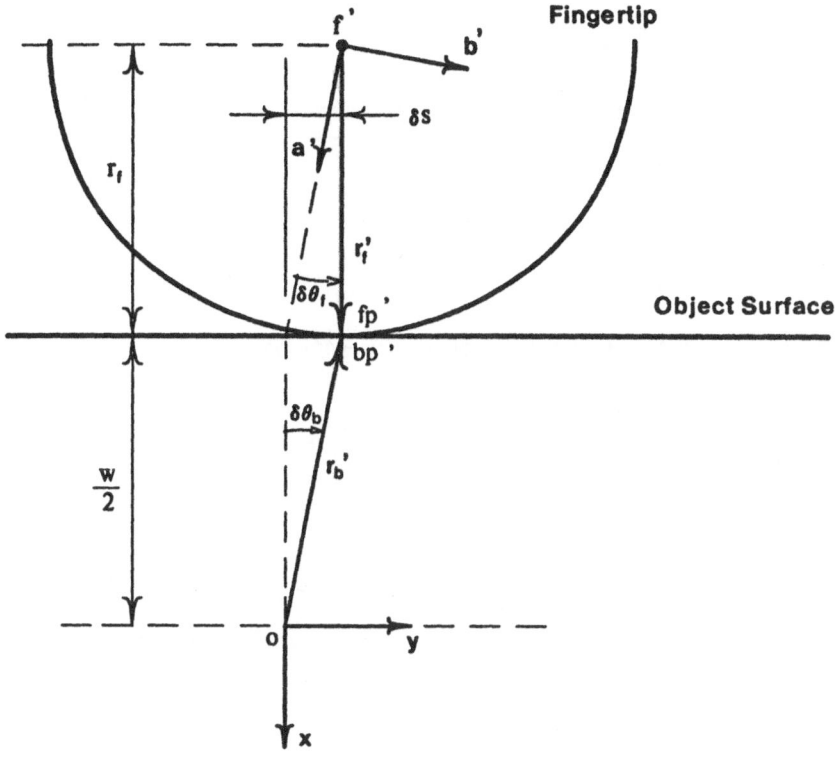

Figure 5-22: Curved fingertip after rolling $\delta\theta$ with respect to object

5.5.3.1 Discussion

The results in the x and y directions are identical to those for the point-contact example but the torque about the z axis has changed. As in the previous example, the expression for torque about the z axis simplifies for the limiting case in which k_c is large compared to k_b. The change in the torque about the z axis reduces to

$$\tfrac{1}{2}(-k_b w^2 + 2fw - 4f r_f)\,d\theta_z$$

In the expression for $\delta g_{\theta z}$ in Table 5-6, if $r_f = \tfrac{w}{2}$ the expression reduces to $-\tfrac{1}{2}\beta w^2 \delta\theta_z$. The physical interpretation is that the translation of the contact point due to rolling of the finger with respect to the object exactly cancels the effect of rotating the object. Thus, for large radii of curvature, $(r_f \geq \tfrac{w}{2})$, the grasp is infinitesimally stable with respect to rotations *regardless* of the stiffness of the fingers.

$$\delta g_x \quad = -k_a\,dx$$

$$\delta g_y \quad = (\alpha f - \beta)\,dy - (f + \tfrac{w}{2}(\alpha f - \beta))\,d\theta z$$

$$\delta g_z \quad = 0$$

$$\delta g_{\theta x} \quad = 0$$

$$\delta g_{\theta y} \quad = 0$$

$$\delta g_{\theta z} \quad = \tfrac{1}{4}\big((\alpha f - \beta)w^2 + 2f(1 - \alpha r_f)w - 4f r_f\big)\,d\theta_z \\ \qquad\quad - \tfrac{1}{2}\big((\alpha f - \beta)w - 2\alpha f r_f\big)\,dy$$

where $\quad \alpha = \dfrac{k_b r_f}{k_b r_f^2 + k_c} \quad$ and $\quad \beta = \dfrac{k_b k_c}{k_b r_f^2 + k_c}$

Table 5-5: Contribution from left finger to δg_b (rounded fingertip)

$$\delta g_x \quad = -2\,k_a\,dx$$

$$\delta g_y \quad = 2(\alpha f - \beta)\,dy$$

$$\delta g_{\theta z} \quad = \tfrac{1}{2}\big((\alpha f - \beta)w^2 + 2f(1 - \alpha r_f)w - 4f r_f\big)\,d\theta_z$$

Table 5-6: Change in g_b due to motions dx, dy, and $d\theta z$ (rounded fingertips)

5.5.4 Very Soft Fingers

For contacts with soft fingers a combined compliance matrix is established for the finger and fingertip as in Section 5.4.3. The combined compliance matrix is shown in Appendix A.5 for a square fingertip. In the matrix, k_p is the elastic stiffness of the fingertip in compression. Since the shear modulus, G, of rubber-like materials is generally about one third the compression modulus, E, the shear stiffness can be written using equations (5.32) and (5.33) as $\tfrac{1}{3} k_p$. From equations (5.34)-(5.36), the bending and torsional stiffnesses are approximately $B k_p$ and $\tfrac{2}{3} B k_p$, where B is equal to one-twelfth the contact area.

The restoring force at the contact is $\delta g_{bp} = [Kbp]d_{bp}$. The restoring forces in joint coordinates are given by $\delta g_q = [Jfq]^t \delta g_{bp}$ and the corresponding motions in joint coordinates are given by $d_q = [Kq]^{-1}\delta d_q$. The motions are then expressed in fingertip coordinates as $d_{fp} = [Jfq]d_q$.

$$\delta g_x = -\frac{k_b k_p}{k_b + k_p}\, dx$$

$$\delta g_y = \frac{-2k_p k_b (k_c - fr_f)}{k_b k_p r_f^2 + k_c k_p + 3k_b k_c}\, dy$$

$$\delta g_z = 0$$

$$\delta g_{\theta x} = 0$$

$$\delta g_{\theta y} = 0$$

$$\delta g_{\theta z} = \frac{k_b k_p (fr_f - k_c)w^2 + 2fw k_p(k_b r_f^2 + k_c)}{2(k_b k_p r_f^2 + k_c k_p + 3k_b k_c)}\, d\theta z$$

Table 5-7: Change in δg_b for small contact area (soft fingertips)

As in point contact and rolling contact, comparison between d_{bp} and d_{fp} determines the relative motion between the finger and the object, which appears in the differential jacobian $\Delta[Jf]^{-t}$.

The net change in g_{bp} is obtained by summing the restoring forces and the forces due to the change in geometry.

$$\delta g_{bp} = [Kbp]d_{bp} + \Delta[Jf]^{-t}g_f$$

5.5.4.1 Discussion

The general expression for Δg_b is lengthy, but it is simplified considerably for the limiting cases in which the contact area is very small, or very large. To further simplify the algebra in the following results, the finger joint stiffnesses in the a and b directions have been set equal so that $k_a = k_b$.

For a small contact area, $B \to 0$ and the bending and torsional stiffnesses become negligible in comparison to the shear and compressive stiffnesses. For two fingers, the final results are given in Table 5-7. If it is further assumed that $k_p \gg k_b$, as is usually the case, it can be shown that the results for Δg_b become identical to those obtained in the point contact case.

For the case when the contact area is large, the bending and torsional stiffnesses become infinite. If it is again assumed that $k_p \gg k_b$, the problem reduces to that of a finger glued to the surface of the object and Δg_b is given in Table 5-8.

$$\delta g_x \quad = -2 k_b dx$$

$$\delta g_y \quad = -2 k_b dy$$

$$\delta g_z \quad = 0$$

$$\delta g_{\theta x} \quad = 0$$

$$\delta g_{\theta y} \quad = 0$$

$$\delta g_{\theta z} \quad = -2(k_c + (\tfrac{w}{2} + r_f)^2 k_b) d\theta_z$$

Table 5-8: Change in δg_b for large contact area (soft fingertips)

5.6 Summary

In Section 5.2 a procedure was listed for discovering the properties of a grip by moving the object slightly, observing the resulting finger motions and determining the changes in the forces on the object. The grip properties of stiffness, resistance to slipping and infinitesimal stability were introduced and it was shown that such properties could be used to compare grips. For specific tasks, one could then choose, for example, the grip that would be stiffest with respect to rotations or the grip that would resist the largest vertical force before slipping occurred.

Two-dimensional examples with point-contact fingers were used to demonstrate how the grip properties depended on finger stiffness, finger arrangement and gripping forces. In later sections a more complete three-dimensional analysis was developed. In the final example of Section 5.2, the instability of a coin held on-edge between two fingers was discussed, using the simplifying example of a rectangle held between two pointed fingers. When the rectangle was rotated slightly, the finger stiffnesses produced restoring forces that tended to stabilize the grip, but the differential change in geometry resulting from the rotation allowed the grasp forces to become unstable. The stability of the grip was a function of the finger stiffness, the length of the rectangle, and the magnitude of the initial grasping forces. Interestingly, the grip became less stable as the gripping forces were increased. Thus, while an increase in the gripping forces may make the fingers more resistant to slipping, it does not always make the grip more secure.

The coin example also uncovered a limitation of the point-contact assumption used in previous analyses. With pointed fingers, a coin is *less* stable if held by fingers pressing against the two faces than if held on edge. For human fingers, this is obviously not the case. A more accurate model of the finger/object interaction (one that accounts for the deformation and rolling of the fingertips) explains why. Such a model is developed in Section 5.4. First, however, it is useful to establish a more general framework for determining the force/deflection relations of a grasp.

For three-dimensional problems it becomes convenient to use matrices to describe the grip. The matrix equations are developed in Sections 5.3.1, 5.3.3, and 5.3.5. When the procedure of Section 5.2 is applied to general, three-dimensional problems, the results depend on the number of degrees of freedom of the contact and the finger. For an arbitrary motion of the object, the finger motion can be classified as under determined, exactly determined or over determined. Different solutions are discussed for each case.

Section 5.4 took a closer look at the interactions between different kinds of fingertips and the object. The characteristics of pointed, curved, and soft fingers were compared. The different characteristics are reviewed below, and summarized in Table 5-9.

In Section 5.4.2, it was shown that the rolling of curved fingers causes the contact area to shift with respect to the object. This adds a new term to the differential change in the geometry of the grasp — one that may help to stabilize it. As expected, when the radius of curvature of a curved finger becomes very small, the movement of the contact point becomes negligible and the contact behaves like a point-contact with friction. As the radius of curvature becomes very large, the finger approaches the limiting case of a flat-tipped finger having a planar contact with friction.

Fingertips	Kinematic conditions	Friction conditions
(1) f_n f_t **pointed**	Point contact with friction. Translational motions and forces are transmitted, but rotations are not. Finger rotates about contact point which remains fixed on object. $\Delta[Jb]^t = [0]$ $\Delta[Jf]^{-t}$: rotation terms	Force tangent to object surface limited by Coulomb friction law $f_t \leq \mu f_n$
(2) f_n r_f f_t **curved**	Only translational forces and motions transmitted. Contact point moves as finger rolls. Approaches case (3) for $r_f \rightarrow \infty$ and case (1) for $r_f \rightarrow 0$ $\Delta[Jb]^t$: translation terms $\Delta[Jf]^{-t}$: translation & rotation terms	Force tangent to object surface limited by Coulomb friction law $f_t \leq \mu f_n$
(3) **flat**	Planar contact with friction. Translational and rotational forces and motions transmitted. No relative motion without slipping. $\Delta[Jb]^t = [0]$ $\Delta[Jf]^{-t} = [0]$	Distributed pressure and shear tractions allow transmission of forces and moments in plane of contact.
(4) **very soft**	Add elastic fingertip compliance to finger compliance. Contact forces produce relative motion. Approaches case (1) for $A \rightarrow 0$ and case (3) for $A \rightarrow \infty$ $\Delta[Jb]^t = [0]$ $\Delta[Jf]^{-t}$: translation & rotation terms	Uniformly distributed pressure and shear tractions. High (adhesive) friction allows large forces and moments to be transmitted in plane of contact.
(5) r_f **soft, curved**	Elastic coupling + rolling motion. Combine cases (2) and (4). Approaches case (1) for $r_f \rightarrow 0$ and $A \rightarrow 0$ Approaches case (3) for $r_f \rightarrow \infty$ and $A \rightarrow \infty$ $\Delta[Jb]^t$: translation & rotation terms $\Delta[Jf]^{-t}$: translation & rotation terms	Non-uniform pressure distribution and shear tractions permit large forces and small moments to be transmitted in plane of contact.

Table 5-9: Summary of contact models derived in Section 5.4

Fingers also deform, and a model was developed in Section 5.4.3 to investigate the importance of deformation. The model considers a very soft fingertip which conforms and possibly adheres to the object surface. The fingertip compliance is added to the finger joint compliance. As the area of contact becomes small, the fingertip becomes more compliant with respect to rotations than translations and approaches the point-contact model. For large contact areas, the rotational compliance becomes much smaller than the translational compliance and the limiting case of a planar contact with friction is approached.

Fingertips such as those found on the human hand display both rolling and deformation.

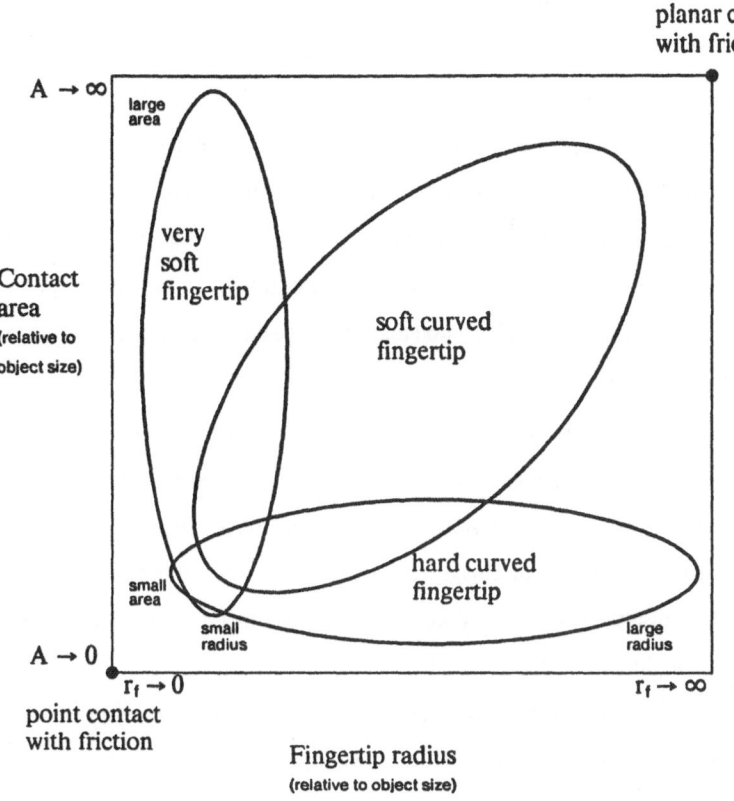

Figure 5-23: Relations between finger models

Section 5.4.4 addressed the properties of a soft, curved fingertip and found that they combined the attributes of the models in Sections 5.4.2 and 5.4.3. As the radius of curvature and the contact area became small, the fingertip could be approximated by a point-contact. For large radii of curvature and large contact areas, the fingertip approached the case of a planar contact with friction.

Figure 5-23 shows the regimes in which the different models developed in Section 5.4 apply, and indicates the limiting cases approached for very large or small radii of curvature and contact areas.

In Section 5.5, some simple examples were used to demonstrate the methods described in Sections 5.3.1-5.3.5 and to illustrate the differences between pointed, curved and soft fingers. When pointed fingertips were used, and only rotations of the object were considered, the problem reduced to the two-dimensional example given in Section 5.2. For curved fingertips, the stability of the grasp increased over the pointed-finger case due to rolling of the fingertips. If the fingertip radii were larger that one half the length of rectangle, the grip became stable with respect to rotational displacements no matter how small the finger stiffnesses were. The relationship between the fingertip radii and the length of the rectangle brings up an important point; the definitions of "large" or "small" radii of curvature and contact areas depend on the size of the object being handled. This is why the point-contact model is reasonable when we hold a basketball or a large box, but not a coin or a matchbox.

Application to hand control

For controls purposes, the small-motion behavior of a grip amounts to a linearized description of the "plant," giving a relationship between displacements of the object and the resulting changes in force. The results show that point-contact finger models and stiffness-based control schemes are not always adequate. If only the stiffnesses of the fingers are considered, a displacement of the object always results in forces that tend to restore the object to its original position. However, a small change in the grip geometry may cause the grasping forces to produce something akin to positive feedback for displacements of the object. For stability, these must be canceled by increasing the grip stiffness in the corresponding directions.

The contribution of the geometric effect varies for pointed, round and soft fingers and its magnitude depends on the relative dimensions of the fingers and the object.

If a stiffness model as used in earlier analyses is not adequate, then what must be done to describe and control the grip? Unfortunately, a three-dimensional analysis of the grip becomes quite involved when finger rolling and deformation are considered. It seems unrealistic to expect a robot or gripper controller to perform a complete analysis in real-time. In the next section, a solution is suggested in which fingertip sensors directly measure the terms that are most difficult to compute.

CHAPTER 6
Natural Examples of Grasping

6.1 The Human Hand

"the instrument of instruments"

(Aristotle)

The hand has been extensively studied in the medical literature and a great deal is now known about how nerves, tendons and muscles control the fingers. It can be an invaluable source of inspiration in designing and controlling grippers, provided we recognize that some attributes of the human hand may not be required,, or even desirable, in a manufacturing environment. There is a tendency to suppose that advanced manufacturing grippers should copy the human hand, since it is the most versatile gripper we know.[7] But the hand has evolved over millions of years as an organ used as much for sensation and communication as for manipulation.

On the other hand, while manufacturing has not been around long enough for the hand to show any adaptation to it, it is certainly true that current manufacturing tools and equipment are designed to be operated by human hands. In fact, the study of ergonomics for manufacturing tasks reveals as much about the functional capabilities of the human hand as it does about the proper design of tools and workpieces [119]. Actually, it has been suggested that homonids used tools over a long enough period for hands and tools to undergo a *mutual* evolution. The first manufactured tools included flint stones, crudely struck together by creatures with hands considerably less flexible than those of *Homo sapiens* [120]. It may be argued that manufacturing processes should evolve along with robots and grippers so that grippers will eventually bear little resemblance to the human hand and the

---

[7] A look at how the human foot resembles the hand reveals the problem with such an assumption; clearly those features that are common to both are not necessarily germane to grasping and fine manipulation.

processes will make no concessions to it. However, if robots are to work with humans in the manufacturing environment they must be able to handle at least some of the same workpieces and tools that people use today. Therefore, a least some of the attributes of human hands may be expected in future grippers.

Some useful properties of the human hand which may be applied to the design of robotic hands are reviewed below.

6.1.1 Conformability

The hand displays both large and small scale conformability. The hand is curved at rest, ready to conform to the exterior surfaces of objects. (We rarely encounter an object that has no convex surfaces for gripping.) The fingers curl about objects, encircling them with little energy expenditure [89]. As the grasp tightens, deformation takes place on a more localized scale as the tissues of the finger pads and the palm adapt to variations of the object surface. The finger tissues are visco-elastic; they adapt gradually and remain deformed after the grasping force is relieved. The imprint from a rope or a drawer handle remains visible in the skin for a couple of minutes after the hand has released its grasp. At a still smaller scale, it has been suggested that the papillary ridges, which are responsible for fingerprints, improve the adhesion of the hand on rough surfaces by interlocking with small projections [121]. Current experiments [122] suggest that the ridges also work like treads on a tire to reduce the buildup of hydrodynamic films when handling smooth, nonporous objects. Without them, sweat, oils and environmental moisture would make such objects extremely slippery. The result of the large and small scale conformability in the hand is that friction and passive restraints contribute most of the required force balance for gripping, with a resulting economy in muscular energy.

6.1.2 Muscles

The finger muscles can be grouped into intrinsic muscles within the hand and extrinsic muscles in the forearm. The former act directly upon the fingers while the latter control the fingers via tendons routed through the wrist. It was once suggested that the intrinsic muscles were used in light tasks requiring precise control while the extrinsic muscles are used in heavy tasks, but this has been shown to be an oversimplification [123]. It does appear, however, that the intrinsic muscles primarily help to stabilize the fingers and to provide fine manipulation, while the extrinsic muscles provide most of the power when the hand closes about a heavy object. Those of us who have taken piano lessons may remember being admonished to curve our fingers while playing softly, so that the little muscles in the fingers would be exercised, rather than striking the keys with stiff digits powered by the muscles in our forearms.

6.1.3 Hand/Wrist Interaction

The human hand and wrist comprise some 27 bones, 48 muscles and 22 degrees of freedom [123, 124, 125]. Fourteen of the bones, or only about half, make up the fingers beyond the knuckles. The remainder are found in the palm and at the wrist joint. In fact, it is difficult to draw a firm line that separates the bones of the fingers from those of the wrist. To look at a skeleton of the hand, one would think that the five metacarpal bones were clearly parts of the fingers. They seem to be distinct, articulated links connecting the knuckles to the base of the hand. In the living hand, however, they form the palm, interconnected by flesh, muscles and ligaments. Much of the versatility of the human hand is due to the wrist and arm that support it. Few tasks are performed by the hand muscles alone since, even if the wrist and arm do not move, they are required to maintain a stable support. As soon as we pick up an object, the muscles in the hand, wrist, arm and torso all become stiffer in order to bear the load. The degree of muscular coupling between the hand and the wrist is apparent in everyday activities:

> "When a fist is made, the wrist must be fixed in dorsiflexion so that the fingers can grip. If not, the wrist falls into palmar flexion and strength of grip is lost. Clinically, the severance of radial wrist extensors results in a loss of 50 percent of grasping power." [126]

6.1.4 Finger Coupling and Specialization

The image of a gripper as a collection of several little manipulators at the tip of an arm has little in common with the human hand. Human fingers are by no means independent but are coupled, elastically, in a way that simplifies prehension — reducing the order of the control problem and providing mechanical stability. Stabilization of the hand is partly active (through opposing muscles) and partly passive (through tendons, sheaths and connective tissues) [123].

The thumb is manifestly the strongest and the most independent of the fingers. The flexibility and controllability of the thumb are among the major features that distinguish the human hand from the hands of primates discussed in Section 6.2. The next most independent finger is the fifth. Loss of the fifth finger is a major handicap for handling tasks, since the fifth finger contributes greatly to the stability of the grip when holding a heavy object. In fact, for a workman, loss of the fifth finger is considered to be more serious than loss of the second, third or fourth [127].

6.1.5 Grasps

The human hand is capable of a variety of basic grasps which are discussed both in the medical literature [126, 89] and in papers on versatile artificial hands [124, 69]. Much of the robotic literature mentions the six grasps defined by Schlesinger in 1919 [128] and summarized by Taylor and Schwarz [129]: the cylindrical grasp, the tip grasp, the hook grasp, the palmar grasp, the spherical grasp and the lateral grasp. However, in a metal working cell, people do not commonly use all six of the above grasps. In addition, they use grasps that are not included in the above list, or perhaps are hybrids of the six basic grasps.

Figures 6-1-6-10 illustrate grasps used by a machinist working at a lathe and a milling machine in a small machine shop. The accompanying text explaining when the grip is used, whether motions are made with the fingers or wrist, and how forces are sensed, represents a consensus among several machinists and the author.

Figure 6-1: Power grasp, with fingers and thumb curled about a heavy object

Power grasp

Figure 6-1 shows the power grasp used in working with heavy tools such as hammers and wrenches. The fingers and thumb wrap tightly about the tool handle. Swinging motions are made with the arm and fine motions (to correct the orientation, or aim of the hammer blow) are made with the wrist. The fingers are passive.

Cylindrical grasp

Figure 6-2 shows the machinist using what Schlesinger calls the "cylindrical grasp" to load a heavy billet into the chuck of a lathe. The fingers and palm wrap around the heavy, convex part. Friction between the fingers and the part accounts for much of the grasping force balance. The chuck-loading task is a heavy peg-in-hole assembly, and is representative of operations for which the robotic wrist in Chapter 4 was originally designed. The arm brings the billet toward the chuck and provides support while the wrist makes small accommodations that align the billet so that it can be inserted.

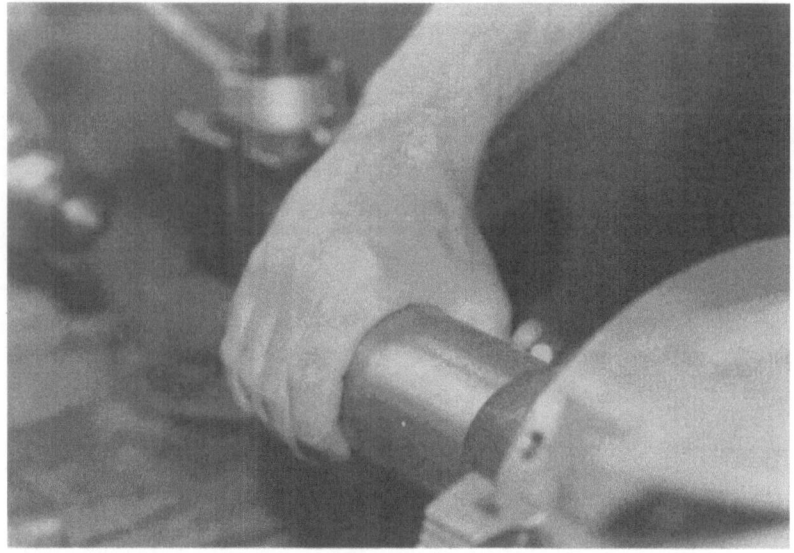

Figure 6-2: Cylindrical grasp for a heavy, convex workpiece

Five-fingertip grasp

The shape of the workpiece and the operation illustrated in Figure 6-3 are almost identical to those of Figure 6-2. However, since the cylindrical billet is now hollow and much lighter, the machinist holds the cylinder by one end, using a grasp in which the tips of all five fingers press radially inward. Both the wrist and the fingers can be used to make the fine accommodations required for insertion. The wrist is used especially for rolling the cylinder about its axis, while the fingers move in and out from the palm to tilt the cylinder. The tips of the fingers sense forces and vibrations.

Modified hook grasp

Figure 6-4 shows a variation of the hook grasp in which the thumb runs along the tool handle for extra control. As in the standard hook grasp used in carrying a suitcase, the main force is a pull applied by the arm. Powerful extrinsic muscles, connected to the fingers by flexor tendons, prevent the fingers from uncurling.

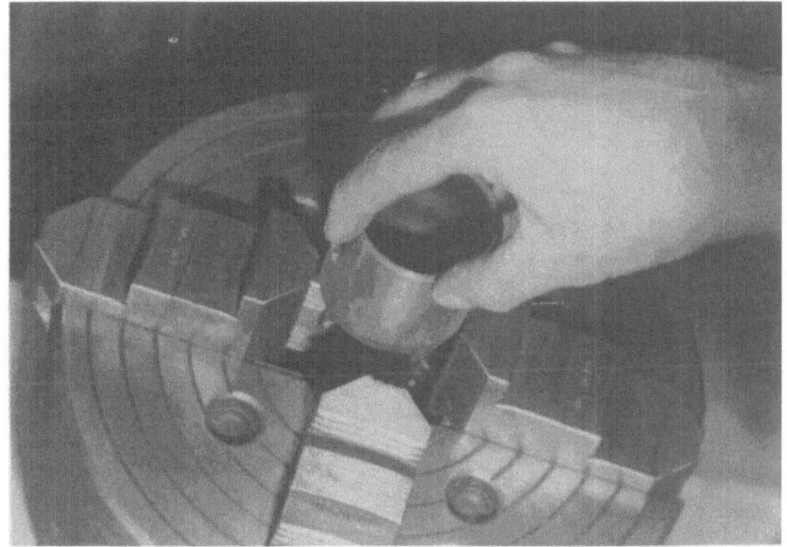

Figure 6-3: Five-fingertip grasp for light cylindrical objects

Modified cylindrical grasp

Figure 6-5 shows a variation of the cylindrical grasp, used when forces are light and a higher degree of sensitivity is desired than in Figures 6-1, 6-2 or 6-4. The fingers wrap gently about a small pneumatic grinder, with the index finger extended (as in the *modified hook grasp* above) for extra control and sensitivity to grinding forces and vibrations. Since the fingers press more gently than in the cylindrical grasp of Figure 6-2, they are more sensitive to small changes in force and vibration. The fingers do not manipulate the tool, but they do actively adjust the stiffness of the grasp.

Modified precision grasp

Figure 6-6 shows a grasp used in assembling fixture components. The grip contains elements of the "palmar grasp" and the "tip grasp" described by Schlesinger, except that the workpiece is too large for the thumb and index finger to be brought close together. Small motions are made with the fingers and larger motions with the wrist.

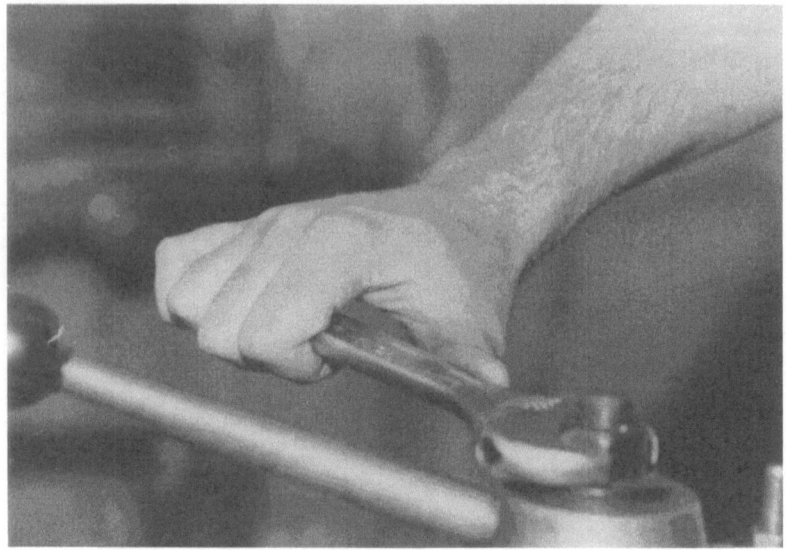

Figure 6-4: Modified hook grasp with thumb along tool

Three-finger precision grip

The hand in the left of Figure 6-7 illustrates a precision grasp in which the tips of the thumb, index and middle fingers hold a small bolt and start to engage it with the thread of a tapped hole. The fingers tilt the bolt, aligning it with the axis of the hole, while rotations about the axis of the bolt are achieved primarily with the wrist. As soon as the thread of the bolt engages the thread of the hole, the machinist switches to the two fingered grasp below, so that he can rapidly twist the bolt with his fingers.

Two-finger precision grip

Figure 6-8 illustrates a precision grip essentially identical to the "palmar grasp" described by Schlesinger. The thumb and index finger roll against the part or tool to twist it rapidly. The grip is often used when the tip of the part or tool is touching another object. The contact becomes a third finger-like constraint upon the object, which permits relative rotation, but no translation, between the two objects.

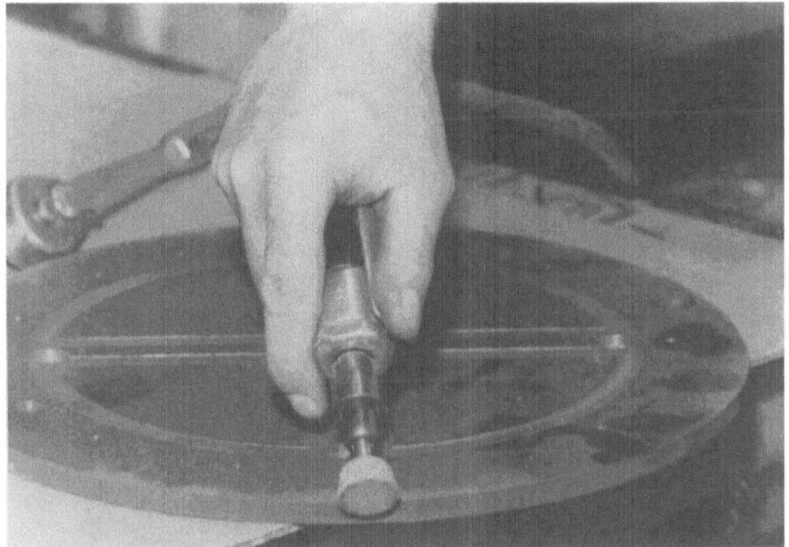

Figure 6-5:   Modified cylindrical grasp with partial fingertip prehension

Lateral pinch

In figure 6-9 the hand adopts the "lateral pinch" described by Schlesinger.   The machinist is using a small scraper to remove burrs from the inside of a pipe.   The grip is achieved primarily with the thumb and the side of the index finger, but the remaining fingers add a degree of support.   The grip is used when the object is small and the task-induced forces are high, so that a powerful grip is required.   Thus, the lateral pinch becomes the grip of choice for turning a key in a lock or for shooting a watermelon seed.

A complex grip

Figure 6-10 illustrates some of the remarkable versatility of the human hand, and points out that a person often uses two hands in close cooperation.   The hand on the right holds a part in modified wrap grip.   The fingers of the hand on the left curl about a scraper in a hook grasp while the thumb presses against the work piece, stabilizing the hand and linking it with the hand on the right.

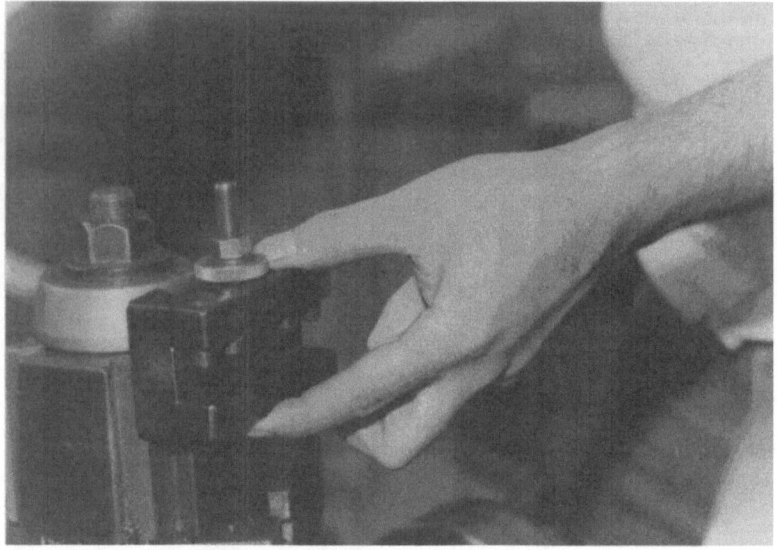

Figure 6-6: Modified precision grasp for objects that are not small with respect
to the hand

6.1.6 Sensation and Control

Our hands are endowed with an enormously sophisticated sensory system. The degree of
complexity is revealed in the number and variety of nerves in the hand and in the fraction of
the brain devoted to processing information from the hand and controlling it.

> "Large areas of the brain coordinate motion in the hand; the area of the
> cortex devoted to the motor coordination of the arms, legs, and trunk is
> approximately equal to the area specific to the hands alone. The hand can detect
> changes in shape, touch, texture, temperature, and movement. One-fourth of all
> touch receptors (Pacinian corpuscles) of the body are located in the hands... "
> [125]

Part of the wealth of sensors is due to the hand's role as an exploratory device. Unlike the
other sensory organs, the hand can engage in active sensing, moving toward a new object,
palpating and manipulating.

The sensors in the hand may be grouped into categories of superficial and deep sensors.

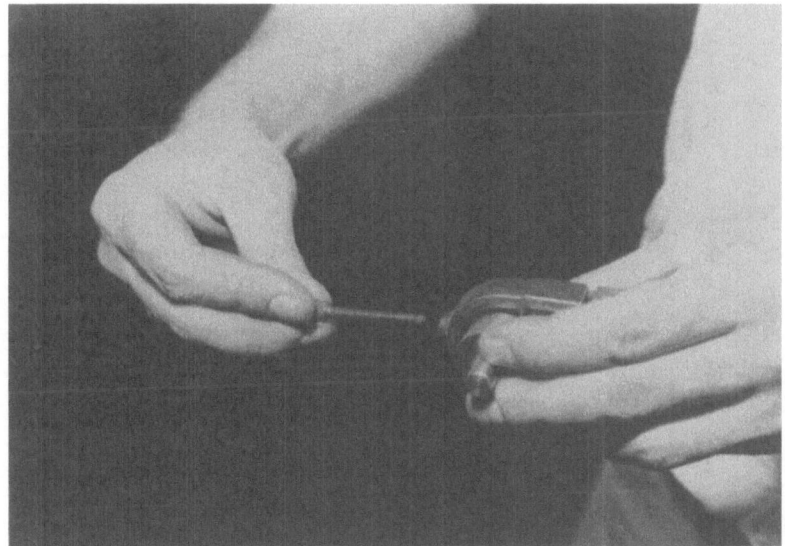

Figure 6-7: Three-finger precision grip using thumb, index and middle fingers

Figure 6-11 shows the arrangement of several of skin sensors. The superficial sensors include the Pacinian corpuscles which are most sensitive to vibration or localized accelerations; Merkel discs, which respond to skin deformation; hair follicle receptors, which respond to a light touch and to hair displacement; and Ruffini endings, which are sensitive to skin displacement and thermal stimuli [93]. The fingertips have an especially rich concentration of Pacinian corpuscles and are also equipped with Meissner corpuscles which sense shear and deformation of the skin. In addition to these mechanical sensors, the hands have a variety of thermal sensors and nocioreceptors which respond rapidly to potentially damaging stimuli.

The deep sensors include muscle spindle receptors, and tendon sensors [93, 131, 132] as shown in Figure 6-13. These provide information for controlling the hand through both low-level and high-level feedback loops.

At the lowest level of control there is the stretch response of opposed muscles. This response accounts for the well known knee-jerk reflex when a physician hits a patient's knee with a rubber hammer. As the tendon is stretched, signals from the stretch receptor travel to

Figure 6-8: Two-finger precision grip, or "palmar grip"

the spinal cord and trigger automatic motor responses to counteract the increase in length. The result is a low-level servo-system that allows a pair of opposed muscles to automatically adjust to disturbances in loading. Signals from the stretch receptors eventually find their way to the motor cortex, but apparently do not contribute to a conscious awareness of muscle stretch [133], (by contrast, we are consciously aware of signals from the tactile sensors in our skin). When we wish to move our limbs, we direct a sequence of signals to the motor neurons which trigger muscular responses. The muscular reponses, in turn, trigger the stretch reflex, which stabilizes the motion of the limb.

With low-level and high-level feedback loops, the hand can be said to have a certain amount of distributed control, although not as much as is found in primitive animals. In fact, it may be indicative of the versatility and general-purpose nature of the human hand that comparatively few low-level reflexes are present.

The reflexes of the hand have been studied as an aid in monitoring the development of infants [134, 135]. Figure 6-12 illustrates several infant grasping reflexes. The best known of

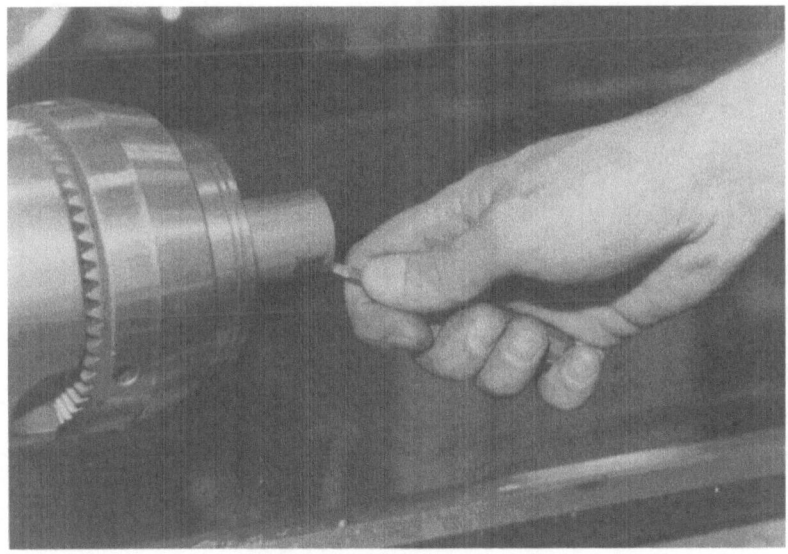

Figure 6-9: Lateral pinch grip

the hand reflexes is the prehensile reflex in which an infant grasps an object placed in the palm and tightens its grip if one attempts to withdraw the object. This reflex diminishes as the infant learns to control the hand in purposeful movement. A more integrated reflex has also been discovered in which

> "... a light stationary or moving contact stimulus to any part of the hand caused an orientation of the hand into a position of readiness for a series of light palpating or groping movements which were then followed by a final grasping of the object." [134]

The exploratory reflex develops over a period of several months and occurs independently of visual stimuli. Acting in opposition to the grasping and exploratory reflexes is an avoidance response in which the hand and arm retract from a moving stimulus, especially if the stimulus is applied to the back of the hand. The reflexes must be balanced for facile hand control. People with a disorder in which one or more of the reflexes are absent show a marked difficulty in acquiring and manipulating objects with the hand [134].

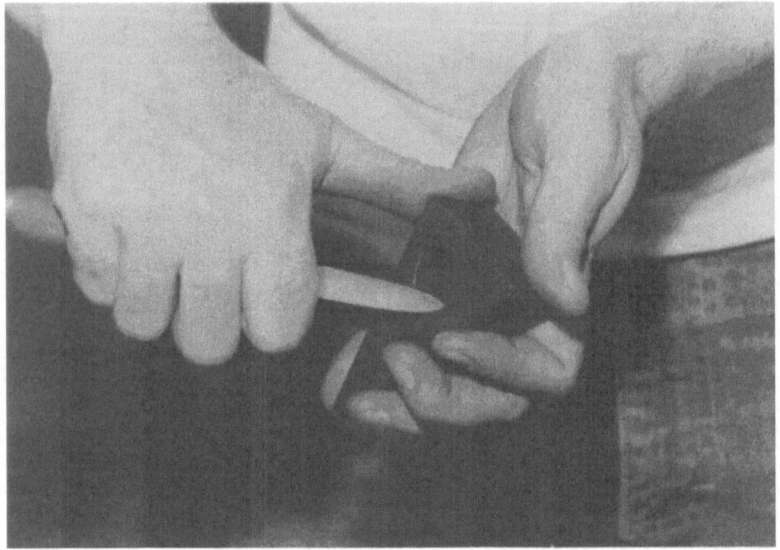

Figure 6-10: A complex, two-handed grip

6.2 Other Natural Examples

Almost any conceivable approach to grasping and manipulation can be found in nature. Most such approaches are highly specialized adaptations to a particular set of tasks. The approaches include pincers, mouths, suction cups, sticky secretions, claws, beaks and flexible members such as trunks or tentacles that curl around an object,

Among the simplest grippers are the pincers of insects and the beaks of birds. These also have the most in common with today's industrial grippers. Insect physiology reveals numerous ways of reducing the size and complexity of the muscular and neuronal sytem for a gripper. For example, grasping requires strength and control only in closing the gripper; very rarely is power or precision needed in opening. Consequently, certain insects and arthropods have large muscles for closing their grippers, but none for opening. Opening is accomplished by the elasticity of the cuticle and by an increase in blood pressure within the limb. A similar though less pronounced simplification is found in more sophisticated

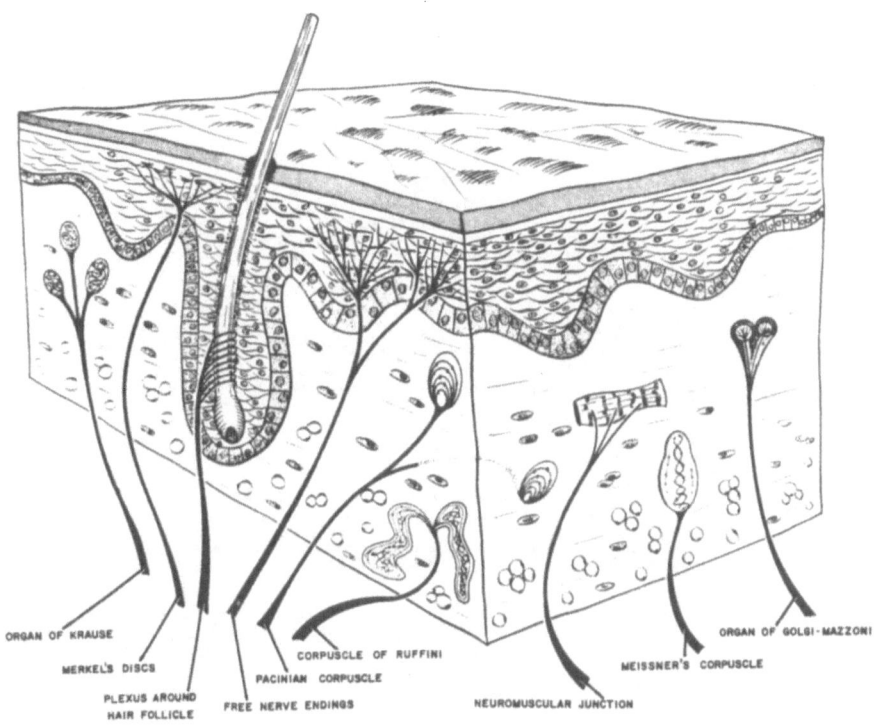

ORGAN OF KRAUSE

MERKEL'S DISCS

CORPUSCLE OF RUFFINI

PLEXUS AROUND
HAIR FOLLICLE

PACINIAN CORPUSCLE

FREE NERVE ENDINGS

NEUROMUSCULAR JUNCTION

MEISSNER'S CORPUSCLE

ORGAN OF GOLGI-MAZZONI

Figure 6-11: Some of the sensors in the skin of a hand (from [130], reprinted with permission from W.B. Saunders Co. and McGraw-Hill Inc.)

grippers, including the human hand. There may be muscles, nerves and tendons for opening, but they are generally fewer and smaller than those provided for closing.

Beaks and pincers lack the tactile information that a hand possesses. Only forces and torques upon the gripper can be measured and consequently, other sources of information, such as vision, may be required for agile manipulation. "In the desensitized hand, vision becomes an indispensable compensatory mechanism." [123] A similar effect has been observed in experiments in which people were asked to assemble objects, with and without blindfolds, and with varying handicaps designed to reduce the amount of tactile information: "Not surprisingly, the impact of visual information is largest when the level of tactile information is lowest, suggesting that to some extent, these two senses can substitute for one another." [136]

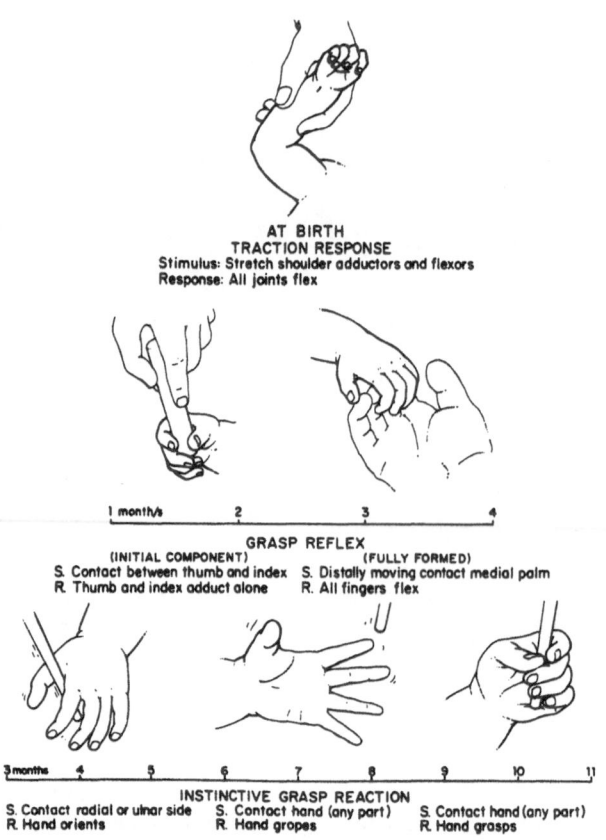

AT BIRTH
TRACTION RESPONSE
Stimulus: Stretch shoulder adductors and flexors
Response: All joints flex

| 1 month/s | 2 | 3 | 4 |

GRASP REFLEX
(INITIAL COMPONENT)   (FULLY FORMED)
S. Contact between thumb and index   S. Distally moving contact medial palm
R. Thumb and index adduct alone   R. All fingers flex

| 3 months | 4 | 5 | 6 | 7 | 8 | 9 | 10 | 11 |

INSTINCTIVE GRASP REACTION
S. Contact radial or ulnar side   S. Contact hand (any part)   S. Contact hand (any part)
R. Hand orients   R. Hand gropes   R. Hand grasps

Figure 6-12: Automatic grasping responses in infants (from [134], reprinted with permission from The Pergamon Press Ltd.)

Closest to the human hand, are the hands of apes and monkeys. Generally, monkeys have a limited ability to bring the thumb into opposition with the index finger. The simian hand is specialized for the power grasp. Increasingly intelligent primates have increasingly flexible hands with larger pads of sensitive skin on the palm and fingers. It has been suggested that the differences between the hands of monkeys and the hands of such animals as squirrels (both of which are agile arboreal animals) is the need for monkeys to grasp branches that are smaller than the hand, while squirrels run along branches and tree trunks that are large compared to their hands [121]. In the first case the power grasp, in which the fingers curl about the branch, is most effective. In the second, there is no possibility of

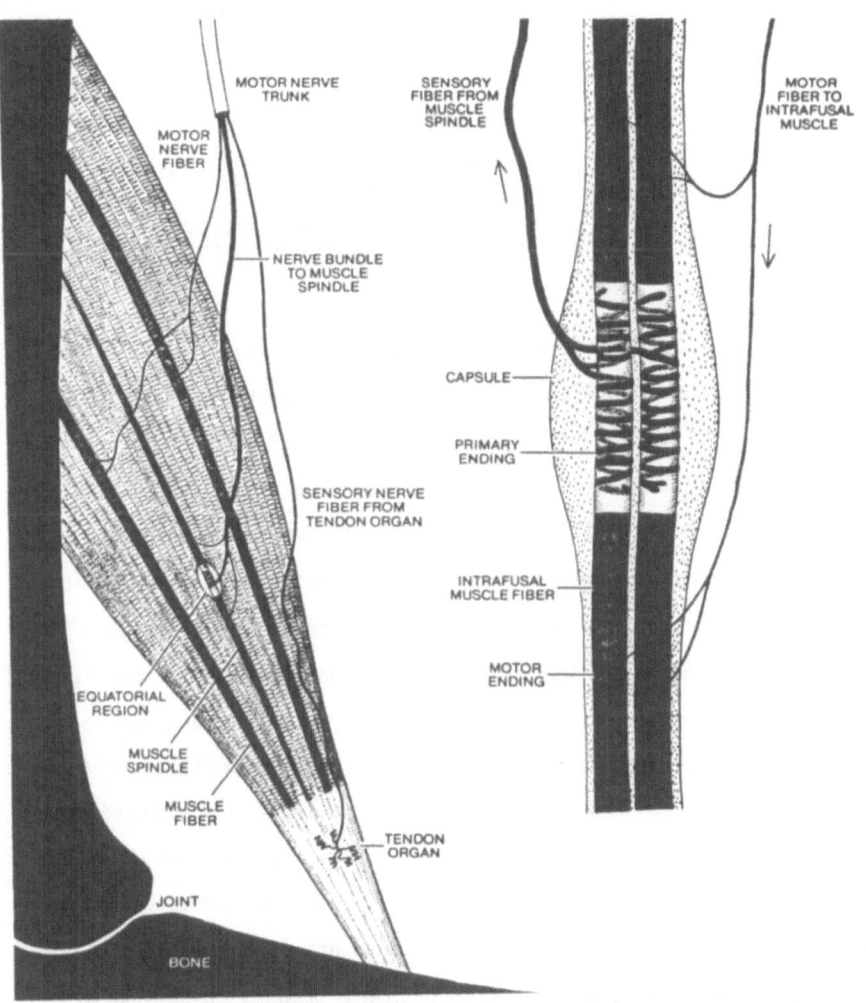

Figure 6-13: Typical arrangement of muscle receptors (from [133], reprinted with permission from The Scientific American)

encircling the branch, and claws are used. When a squirrel needs to hold a smaller object, a nut for example, it traps the object between two hands. The important factor then, is the relative size of the object to be grasped. The importance of relative object size also surfaces in the grasping calculations of Chapter 5.

CHAPTER 7
Designing Hands and
Wrists for Manufacturing

A recurrent theme in this thesis is that many manufacturing tasks can be broken into fine motion and gross motion subtasks, and therefore it is appropriate to develop a manipulation system consisting of separate "modules." In the case of assembly, the operation can be decomposed into gross motions that move the peg in the direction of the hole, and fine accommodations about the central axis of the hole. If a robot is to assemble parts, it suffices to have an arm with the required reach and speed to pick the parts up, move them roughly into position, and push them toward each other, while a compliant RCC wrist takes care of the fine accommodations and keeps the parts from jamming. In other tasks, such as grinding, the fine and gross motion elements are not entirely independent; a change in the velocity along the surface changes the amount of material removed and alters the apparent "hardness" of the surface. Still, it is possible to use a separate arm, wrist and hand provided that they communicate with each other. If the tasks displayed a large amount of interdependence between fine and gross motions, it would become more convenient to treat the wrist and hand as the last few joints at the end of an arm. In the present case, it suffices to work in cartesian coordinates, transmitting deflections or forces sensed at the wrist and hand, and locations reported by the robot. It is not necessary to directly couple, say, the third wrist sensor with the fourth arm actuator.

In the above scheme, the hand and wrist become special-purpose manipulation devices. Based on the experimental and theoretical results in Chapters 4 and 5, particular actuators, sensors, stiffness properties, geometries and gripping surfaces are considered below that take advantage of the special-purpose nature of hands and wrists.

7.1 Wrist Design

Passive and active wrist functions

For the most part, the wrist in Chapter 4 does not add new motions to those of the robot arm but modifies the motions transmitted through it. A servo controlled wrist could make small, precise motions and could be used for such tasks as positioning a machined part in a fixture while the fixture was being tightened. The wrist in Chapter 4 would become a servoed device if each sphere were given a separate, dynamically controlled pressure source instead of sharing a single cylinder adjusted by a stepping motor. With this modification, each sphere becomes a small bladder-type hydraulic actuator. The design would be inherently different from those described elsewhere [41, 40] because it would remain a compliant RCC device with a monotonically increasing force/deflection curve. The passive properties of the wrist would still be able to accomplish assembly and contour following tasks — even in the absence of active control. The result would be a servo controlled wrist with built-in mechanical "reflexes." The wrist controller would be free to modify such reflexes, and to add actions to them. As seen in Chapter 6, biological systems also have passive mechanisms which simplify their control.

Low power, high force

A wrist should be designed as a low power, high torque device. The forces and moments transmitted through the wrist may be nearly as large as those transmitted through the robot arm, but since the wrist never has to move very far or fast, the power requirements are small. This has implications for the drive mechanism used in the wrist. Hydraulic actuators are suited for direct-drive applications involving large forces and low speeds. Electric motors should be geared down if used in a wrist.

Resolution

Since the wrist has a small working envelope it has the potential to be much more accurate than a small robot. The sensors in the wrist work over short distances and the limiting resolution obtained with, say, a 12 bit analog to digital converter is much better than it is for a robot arm. For the wrist in Chapter 4, the sensors work over a range of ± 0.125 inches which results in a theoretical resolution of better than 0.0001 inch. In practice, the accuracy of the wrist is closer to 0.001 inch, but this is still better than the most accurate industrial robots can manage. Since the wrist has a single moving upper platform, it is easy to add redundant sensors which reduce the effects of noise in the signal of any single sensor. For example, the wrist in Chapter 4 uses eight sensors for six degrees of freedom.

Compliance

Compliance is of central importance to a wrist. This is partly because with low speeds and small motions, the dynamic equations of the wrist reduce to stiffness equations; impedance and admittance become stiffness and compliance. When a passive, two-fingered industrial gripper is used, the wrist becomes the compliant link matching the robot to its external environment. As such, its compliance properties should meet the *average* requirements for manufacturing tasks. For many tasks, including assembly and surface finishing, it is useful if the stiffness matrix can be diagonalized near the tip of the part or tool. In addition, it is useful to increase the stiffness as the tool or workpiece becomes heavier and task-induced forces become larger. Matching the wrist stiffness to the load ensures that the wrist does not deflect excessively under heavy loads, but remains sensitive when loads are light. Another advantage is that the response of the wrist and workpiece to periodic excitations (perhaps from an unbalanced polishing wheel) is more constant than it would be if differences in the masses of tools, grippers and workpieces were not matched by changes in wrist stiffness. The effect can be achieved mechanically (as in the wrist in Chapter 4) in electronic hardware, or in software running on a dedicated microprocessor.

Kinematic design

Given the small working envelope and the high force, low speed requirements of the wrist, it is useful to design the wrist as a parallel kinematic chain, with multiple closed loops. In a serial chain, a small angular error in one of the base joints is magnified into a large positional error at the tip of the chain. In a parallel chain device, the errors in individual links are merely summed. If the wrist has at least $n+1$ actuators for n degrees of freedom it becomes possible to simultaneously control stiffness and position. Actuators can be controlled to work with each other to move the upper platform of the wrist, or against each other, resulting in no motion but an increase in stiffness. In the same way, when we tense the opposed muscles in our arms and wrists, we increase their tonus without moving them. The

RCC devices in [42, 43] and the wrist in Chapter 4, with four axial spheres and eight cables, are examples of parallel linkages.

Friction

Since the wrist does not move far in any direction, the moving parts of the wrist can be made without sliding contacts and without the stick-slip friction that plagues most servo devices. Plastic hinges and flexible metal or plastic elements can replace pinned joints. Diaphragm and bladder cylinders can replace conventional hydraulic cylinders, eliminating pistons and seals. The wrist in Chapter 4 has no sliding friction since the spheres deform and roll to accommodate motions of the upper platform. In a lighter application, Glassman *et al* [137] have developed a micro-manipulator with two degrees of freedom for the tip of an electronic assembly robot. Four magnetic coils position the upper platform of the micro-manipulator, which is suspended on flexible metal strips. Still another possibility is to use solenoids built like loudspeakers, with the moving element suspended in a flexible "spider."

7.2 Hand Design

7.2.1 Grasping *vs* Manipulation

Part of the difficulty in designing an active hand, as opposed to an active wrist, is that it is less a special-purpose device. The manufacturing hand acts in different ways when it grasps parts and manipulates them. Among the most important differences are:

- **Large *vs* small motions**
  Grasping involves conforming to the workpiece, perhaps by placing fingers about the surface or perhaps by surrounding the part in a compliant medium. In either case, the motions are not small compared to the fingers and hand. Thus, bringing the hand into the configuration of a grasp is not a fine-motion operation. Once the grasp is achieved, however, manipulation may consist of fine motions. Admittedly, when people screw a bolt into a tapped hole, or use a small screwdriver to adjust an electronic component, the rotation they impart to the bolt or screwdriver may be large (nearly one full revolution). However, when tools or parts are too heavy to manipulate in this way, people choose a single, firm grip and impart only small motions with the fingers.

- **Finger coupling**

  While the fingers are closing upon an object, they are independent. Their motions should be coordinated, but there is no mechanical coupling between them. However, when an object is manipulated with fingers, it couples their motions. If there are no redundant joints in the fingers, all joints are interdependent. A change in the position or torque at the first joint of a finger affects not only the remaining joints of the same finger, but the joints of other fingers as well.

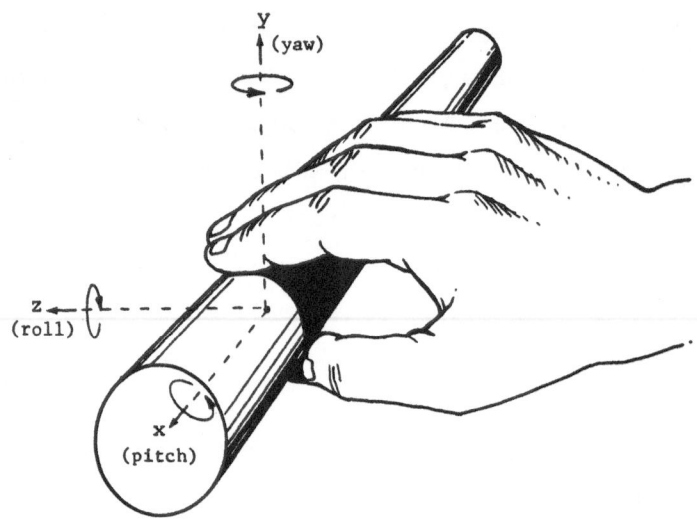

Figure 7-1: Roll, pitch and yaw manipulations with the fingers

Solutions to the different requirements of grasping and fine manipulation are:

1. Design a hand that is sufficiently versatile for grasping and sufficiently precise for manipulation.

2. Design a hand with separate subsystems for grasping and manipulating.

3. Design a hand that only grasps, and leave manipulation to the wrist.

1. A precise and flexible hand

The difficulty is that precision and flexibility are generally antithetic goals. The multi-jointed fingers of a hand give it enormous flexibility in achieving grasp geometries, but controlling the hand becomes a problem in simultaneously and accurately controlling several small, compliant manipulators. By contrast, a wrist can be controlled more easily and precisely than a small robot because it can function in a much more restricted working volume. Still, the goal of a precise and flexible hand is being pursued, and the designs in [77, 76, 81] are significant efforts in this direction.

2. Separate systems for grasping and manipulation

It may be possible to partition the design of the hand into gross and fine motion components for grasping and manipulation, respectively. This would be analogous to the partitioning between the arm and the end-effector, but on a smaller scale. Like the wrist, the fine manipulation elements of the hand could take advantage of linearized kinematics, low power requirements, and small movements. Many of the design concepts discussed above for wrists would apply to the manipulation elements of a hand. The gross, or grasping motion elements would not need to be precisely controlled since they would provide only the basic grasping force and orientation.

Unfortunately, there are two drawbacks to the above scheme. The first is that the scheme adds mechanical complexity to the hand, requiring two sets of actuators and control systems: one for grasping motions and one for fine manipulation. The other drawback is that, like all partitioning schemes, it makes the resulting design less general-purpose. In particular, the scheme does not account for large manipulations with the fingers. As discussed in Chapter 6.1, the human hand reveals limited partitioning between intrinsic muscles in the hand and powerful extrinsic muscles in the forearm. The intrinsic muscles are used more for fine manipulation and the extrinsic muscles more for powerful grasping, but the division is by no means perfect. Frequently, the intrinsic muscles work in conjunction with the extrinsic muscles, providing extra stability and control.

3. Manipulate with the wrist

All the tasks discussed in Chapter 3 could be performed with an active wrist and passive gripper. However, as discussed in Chapter 5, part of the motivation for an active hand is that it can manipulate objects in ways that a wrist, especially a small-motion wrist, cannot. In particular, the wrist cannot rotate an object in pitch or yaw rotations (*see* Figure 7-1) without also translating.

Hand/wrist interaction

A compromise involving aspects of each of the above solutions is for the hand to emphasize those manipulations least well performed by a wrist. To some extent, we do this with our hands. If we wish to rotate a thin cylinder (such as a pencil) about its long axis using our fingers, we rarely hold it so that it points along our fingers and wrist. Instead, we hold it nearly perpendicular to our fingers and wrist and we roll it by moving the fingertips in and out from the palm as in Figure 7-1. If we *must* align the cylinder so that it points along our wrist, it becomes more convenient to rotate the pencil simply by twisting our wrist. Thus, fingers are more important for pitch and yaw rotations than for pure roll rotations.

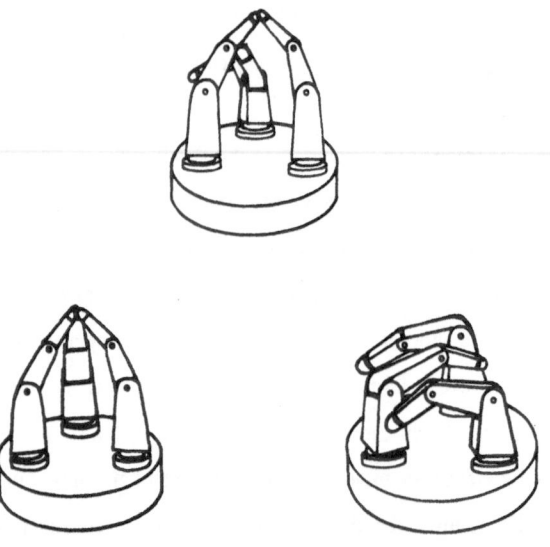

Figure 7-2: Three-fingered hand showing wrap,
and two- and three-fingered pinch grasps

A way to simplify an active hand is to accurately servo only those joints that contribute importantly to finger manipulations. Joints used primarily for grasping, or that duplicate manipulations performed with the wrist, could use a simple control system. For example, Figure 7-2 shows a three-fingered hand with swivel axes at the bases of the fingers. The arrangement is similar to a design by proposed by Skinner [69]) for use on the NASA space shuttle. As shown, the swivel axes are used mainly to switch the hand from a wrap grip, in which two fingers oppose a "thumb," to a precision grip in which two or three fingers point directly toward each other. Since the swivel axes are used primarily for reconfiguring the hand, a precise servo loop is less necessary for them than for the bending joints further along the fingers.

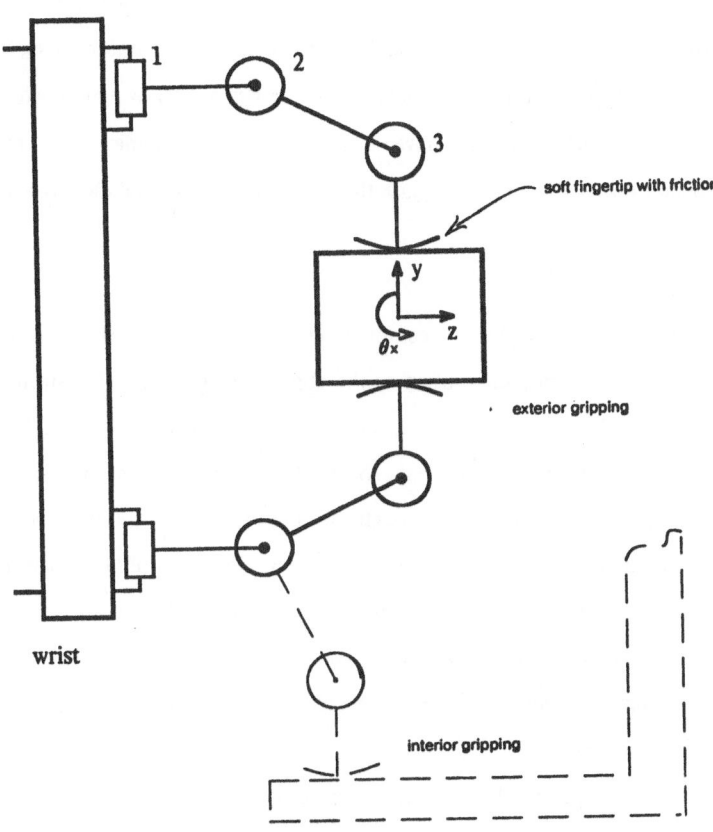

Figure 7-3: Simplified, anthropomorphic two-fingered hand

Why use fingers?

Most of the active hands built to date are anthropomorphic designs with several jointed fingers. But fingers are not required to manipulate grasped objects. The non-anthropomorphic design in [83] can manipulate objects in up to five degrees of freedom (or four instantaneously independent degrees of freedom) using rotating belts. In fact, the advantage of fingers seems to be less for manipulation than for grasping. Until the fingers have closed upon an object, they are open kinematic chains, like manipulator arms. They enjoy the same advantages as manipulators — flexibility and a large working envelope. A two-fingered gripper with enough joints (say, three per finger) to manipulate an object in five degrees of freedom will have a large grasping volume. However, the non-anthropomorphic hand in [83] has just the same grasping capabilities as a conventional parallel-jaw gripper.

Finger sensors

The procedure in Chapter 5 is complex mostly because it is a predictive or open-loop calculation in which only the motion of the object and the physical characteristics of the object and the fingers are assumed to be known. The forward force and displacement relations are relatively simple, but some complication arises in determining how displacements will be transmitted through the contact and how the finger will respond to them. Further difficulty arises in determining how finger stiffnesses, finger motions and grasp forces will interact to change the forces transmitted to the object. The computations would be much simpler if *finger motions* and *contact forces* were available from another source. In practice, humans and animals use sensory information and experience to provide this kind of information.

When we manipulate objects with our fingers, we do not use a kinematic analysis to predict how the forces at our fingertips will change in response to displacements of the object. Instead, we acquire a database of general grip behavior and we use the sensors in our fingers and fingertips to modify our predictions while we work. A similar approach might also be used by a robot, provided the gripper had sufficient sensors to describe the behavior of the grasp. This prompts us to consider what kinds of sensors would be useful. Based on the results of the analysis presented earlier, several types of sensory information are suggested:

- The measurement of normal and shear forces at the fingertips.
  If these can be measured, they do not have to be computed. The shear force can be compared with normal forces and, using information about the friction conditions, predictions can be made concerning how close each finger is to slipping.

- The location of the center of the contact area on the finger.
  Using this information, one can determine how the contact has moved since the last time step, and (by extrapolation), where it will be next. For curved fingers this allows one to track the movement of the contact with respect to the finger and to determine the degree of rolling motion. For fingers that do not roll, it shows the rate at which the finger is sliding against a surface.

- The size, uniformity and general shape of the pressure distribution of the contact area.
  The pressure distribution could be compared with typical profiles for point contacts, curved contacts and soft contacts and an estimate made of how closely the actual contact approaches each of these models.

- Sums and first moments of pressures and shear tractions.
  These allow the forces and moments transmitted through the contact to be determined.

To the above list of fingertip quantities would be added the joint angles and joint torques of the fingers, but already, the list is becoming unrealistic. Even if accurate sensors were available, computing first moments and matching pressure profiles might take the robot just as long as performing the analysis in Chapter 5. Determining such quantities has much in common with feature extraction for grey-scale vision, which is notoriously slow unless performed on special-purpose hardware.

However, even if only the forces and an estimate of the contact size and location were available, the analysis could be simplified. Between these fingertip quantities and the finger joint angles, most of the information needed to describe the grip would be available through forward transformations. The finger joint torques are easily found from the fingertip forces and the fingertip motion is easily determined from the joint angles. An estimate of the contact size would indicate the degree of finger/object coupling and the contact location would allow the finger jacobians to be updated. A small number of fingertip sensors might be sufficient. Recent studies with human beings performing assembly-line tasks

[138] suggest that a sparse array of sensory information (perhaps no more than eight points per fingertip) provides adequate information.

7.3 Control

A fundamental question concerning control for hands and wrists is "How much intelligence is it reasonable to place in the wrist and hand?" As seen in Chapter 6, our own hands are endowed with reflexes that simplify the actions of manipulating objects. Simpler animals, such as primates and rodents with more specialized hands, have more reflexes. The motivation for distributed control and low-level reflexes is that they simplify what is otherwise a formidable problem. Recent investigations on hand control have been concerned with specifying torques and displacements at the finger joints to achieve a desired grasping force, grip stiffness or motion of the object [77, 76, 86]. The problem of controlling the fingers becomes that of controlling several tiny robots at the tip of a robot. The next step is to determine basic reflexes for hand and wrist controllers. Based on the results in Chapters 4 and 5 the following are suggested:

- In general, grip as gently as possible without letting the object slip. A light grip helps to prevent damage to the object and the fingers, reduces the likelihood of instability, and keeps the sensors working near the lower end of their range (where they are often more sensitive).

- Try to match the stiffness of the grip to the requirements of the task. This will simplify the higher level control of the object. In particular:

  o Increase hand and wrist stiffness in proportion to the magnitude of sensed loads.

  o Place the center of compliance near the tip of a part or tool that is being assembled.

  o Be compliant where the sensed stiffness of the environment is high, and stiff where it is low.

All such reflexes would be modified by changes in task description and commands from a supervisory controller.

Above the reflexes in a hierarchy are algorithms for choosing a grip, determining internal

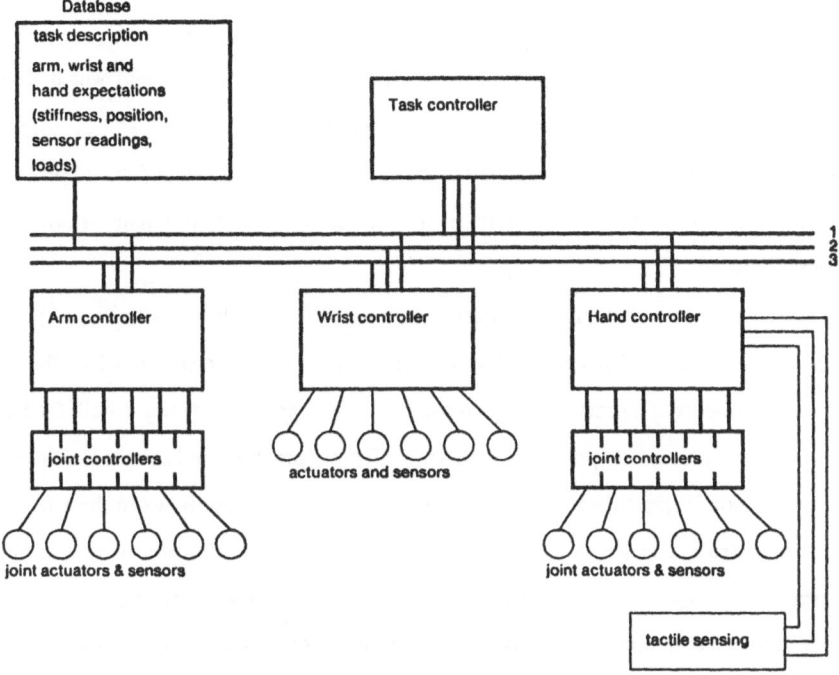

Figure 7-4: Communications scheme for arm-wrist-hand combination

forces to impose on the object and making fine motions in fulfillment of a task. Ideally, choosing and adjusting a grip is something that a robot should be able to do using a combination of analytic methods (including those in Chapter 5), sensory information and some "rules of thumb." The rules are difficult to define, but as we continue to explore the mechanics of gripping and to observe how humans and animals handle objects we can begin to formulate some:

- Spacing the fingers closer together results in a grip that is less stiff with respect to rotations.

- Point contacts are usually less stable than soft or rounded fingers.

- A soft rounded finger slips in rotation before it slips in translation.

Figure 7-4 shows a possible control scheme for a hand-wrist-arm system which would supersede the arrangement in Figure 4-4. The communications lines, 1, 2 and 3, suggest a parallel bus but in fact, serial lines as used in the control system of Chapter 4 would probably be adequate. The important feature is the ability to assume different priorities of communication.

Normally, communications will consist of commands from the task controller, responses from the arm, wrist and hand and data transferred between the controllers and the task database. However, if a programming error results in the hand running into a wall, the hand controller must assume top priority and interrupt all other processes and communications. A similar function is (primitively) fulfilled by the hard wired interrupt between the wrist and the arm controllers in Figure 4-4. With the system in Figure 7-4, the hand controller transmits an emergency statement which the task controller immediately begins to evaluate and for which it determines a response. In the meantime, the arm controller may react to the emergency statement with a withdrawal reflex. The arm would be emulating the pain avoidance response in the limbs of humans and animals. One simple approach is for the arm to retreat to a pre-defined "safe" position until the task controller overrides it with a new strategy.

Control for coupled fine and gross motions

In grinding tasks the position and velocity of the robot must be related to the stiffness and sensory information in the wrist. In Chapter 4 all communication took place through the task controller, but with the system in Figure 7-4 direct communication between the wrist and arm is possible. The task controller would be able to "eavesdrop" on the wrist/arm communication and would remain able to send higher priority commands to either the arm or wrist.

CHAPTER 8
Summary and Conclusions

1. General-purpose robots used in automated manufacturing cells are fast and flexible, but inherently unsuited for fine motion work. Methods are discussed in the literature for improving the accuracy of robots, but a better solution is to endow the robot with a wrist and active hand for fine work. The solution is effective because many manufacturing tasks can be broken into separate fine and gross motion subtasks.

2. Assembly tasks may be broken into compliant accommodations that align the parts to be assembled and larger motions that push the parts together. The directions of the large and fine motions are orthogonal and there is no force/torque coupling between them. Since assembly tasks are not time-dependent, the subtasks do not have to be synchronized. In grinding and surface finishing, the fine and gross motion subtasks are partially coupled. The removal of metal from the workpiece makes these tasks time-dependent and blurs the distinction between directions of free motion and constraint. Changes in robot velocity produce changes in the grinding force and depth of cut.

3. In the experimental work, a fine-motion wrist has been added to a large robot. The remote-center-compliance properties of the wrist permit the robot to accomplish assembly tasks without sensory feedback. For surface finishing tasks, however, the wrist uses sensors and communicates with the robot arm to allow for the coupling in time and force between fine motions of the wrist the gross motions of the arm.

4. A distributed control system, involving a wrist controller, an arm controller and a task controller has been demonstrated in contour following and surface finishing tasks. The communications bandwidth between separate controllers is lower than for a monolithic system, but this is acceptable if

    • The wrist and arm controllers compress sensory information before sending it to the task controller.
    For example, the wrist in Chapter 4 filters signals from its eight sensors and converts them into a vector of translations and rotations in cartesian coordinates.

- The individual controllers are endowed with low-level control actions that reduce the workload of the task controller. For example, the wrist might automatically compensate for static bending loads.

- Efficient, adaptive filtering and prediction algorithms are used to estimate the "state" of the task and the commanded robot trajectory. In typical tracking tasks these methods reduced the number of required steps from 49 to 27 while slightly improving the accuracy of the robot path.

A similar approach is found in biological mechanisms, where slow communications between the limbs and the cerebral cortex are compensated by local spinal and muscular reflexes.

5. An active hand has the potential to increase the versatility of the robot arm and wrist. As a first step, Chapter 5 develops a method for determining the mechanical properties of an active robot hand holding and manipulating an object. When the analysis of Chapter 5 is combined with previous analyses on the kinematics of fingers, the result is an instantaneous or small-motion model of how the grasp will behave in response to task-induced forces and displacements.

6. For a given hand, different grips can be compared in terms of stiffness, stability and resistance to slipping. The results explain, for example, the precarious stability of a coin held on edge between two fingers. Stability is a function of the stiffness of the fingers, the gripping force, the size of the coin with respect to the hand and the finger geometry. Extending the analysis to general three-dimensional problems reveals that:

- While the stiffness of the fingers produces restoring forces, $[Kq]d_q$, that stabilize the grip, small changes in the grip geometry produce changes in the grasping forces, $\Delta[J]^t g_{fp}$, that are of comparable magnitude and may be *destabilizing*.

- If the fingertips are modeled as point-contacts with Coulomb friction, as in previous analyses, the results may be misleading — especially when the object is small compared to the hand and when rubber-like materials are used on the gripping surfaces.

- Rolling and deforming fingertips are examined using linearized models. As a finger rolls on the surface of an object, the relative motion of the contact area increases with the radius of curvature, r, of the fingertip. For a soft finger, the ability of the contact to sustain forces and moments can be expressed as a function of the contact area, A. At extreme values of r and A, the rolling and soft contacts reduce to simpler point or planar contact models:

$$r \rightarrow 0 \qquad \approx \text{point contact with friction}$$
$$r \rightarrow \infty \qquad \approx \text{planar contact with friction}$$
$$A \rightarrow 0 \qquad \approx \text{point contact with friction}$$
$$A \rightarrow \infty \qquad \approx \text{planar contact with friction}$$

Table 5-9 summarizes the properties of different fingertip models.

7. The rolling and deformation of the fingertips considerably complicates the kinematic analysis and the most time consuming terms to compute are the changes in the fingertip forces and moments. However, such terms could also be measured with sensors. Among the most useful quantities to measure are:

   - Normal and shear forces at the fingertips

   - Finger joint angles

   - Contact areas and centroids on the fingertips

8. The analysis supports the notion that a robot can be given grasping rules to help it choose between different grips. Some simple rules that can be argued on mechanical grounds are

   - Grip as gently as possible without permitting the object to slip. This prevents damage to the object, reduces the likelihood of instability and keeps fingertip sensors working near the lower end of their range (where they are often more sensitive).

   - If we don't want a particular finger to slip, we should make it softer than the other fingers.

   - A finger usually starts slipping in rotation before it slips in translation.

   Additional rules are suggested in Section 7.3.

9. The experiments and analyses lead to some general principles for designing active wrists and hands:

   - An active wrist should also have a passive mechanical compliance that matches the *average* needs of assembly and surface-finishing tasks. In particular, it is convenient if the stiffness matrix becomes diagonal near the tip of the workpiece or tool.

   - The hand and wrist should stiffen in response to increasing forces. A higher level task controller is free to modify or override the stiffening reflex. The result is similar to the stretch reflex in the opposed muscles in animals, which automatically compensates for changes in loading.

- Since the wrist has a small working envelope, it should be designed as a low power, high torque device. The actuators and linkages can be designed to flex and bend, without sliding contacts and without stick-slip friction.

- The wrist may be designed as a parallel chain device with redundant actuators and sensors. The sensors operate over small distances and their theoretical resolution is high. For the wrist in Chapter 4, eight sensors measure deflections of the wrist in six degrees of freedom, with a working accuracy of better than 0.001 inch for translations and 0.001 radian for rotations.

- The hand performs different kinds of actions when it grasps and manipulates. To some extent, it is possible to use different mechanisms for grasping and finger manipulation, especially if we concentrate on motions that are not easily made with the wrist. For example, the base of joints of the fingers may be used more to establish the general configuration of the grasp than to manipulate, in which case they do not need to be precisely controlled.

APPENDIX FOR GRASP ANALYSIS

A.1 Matrix Identities

The finger positions and orientations may be expressed with 4x4 homogeneous transformation matrices, $[T]$:

$$[T] \quad = \quad \left[\begin{array}{c|c} A & r \\ \hline 0 & 1 \end{array} \right] \quad = \quad \left[\begin{array}{ccc|c} a_x & b_x & c_x & r_x \\ a_y & b_y & c_y & r_y \\ a_z & b_z & c_z & r_z \\ \hline 0 & 0 & 0 & 1 \end{array} \right]$$

$[A]$ is a 3x3 orthonormal matrix of direction cosines, expressing the orientation of the finger (a,b,c) system of Figure 5-7 in terms of the global (x,y,z) system. r is a vector from the origin of the (x,y,z) system to the origin of the (a,b,c) system. If r_f is the vector in Figure 5-7 from f to fp in (a,b,c) coordinates then $[A]r_f$ gives the same vector in (x,y,z) coordinates. Consequently, the vector from o to fp in Figure 5-7 is $r = r_b - [A] r_f$.

The relationship between two six-element vectors ($d^t = [d_x, \; d_y, \; d_z, \; d_{\theta x}, \; d_{\theta y}, \; d_{\theta z}]$) of differential translations and rotations may be expressed as a 6x6 jacobian.

$$d_{bp} = [Jb] d_b$$

The jacobian is conveniently written in terms of 3x3 partitions:

$$\begin{array}{c} [Jb] \\ (6x6) \end{array} \quad = \quad \left[\begin{array}{c|c} A^t & A^t R^t \\ \hline 0 & A^t \\ (3x3) & (3x3) \end{array} \right]$$

$[A]$ is again a 3x3 matrix of direction cosines. In the above example, $[A]$ expresses the orientation of the (l,m,n) coordinate system at bp in Figure 5-7 with respect to the (x,y,z) system. Since $[A]$ is orthonormal it follows that $[A]^t = [A]^{-1}$.

[R] is an antisymmetric cross-product matrix formed from the elements of a vector r, such that if v is a three-component vector (for example, the three rotational components of d_b) then

$$[R] \; v \; = \; \begin{bmatrix} 0 & -r_z & r_y \\ r_z & 0 & -r_x \\ -r_y & r_x & 0 \end{bmatrix} \begin{bmatrix} v_x \\ v_y \\ v_z \end{bmatrix} \; = \; r \times v$$

Since is [R] is antisymmetric, $[R]^t = -[R]$ and $[R]^t v = v^t [R] = v \times r$.

Given the above identities for [R] and [A] the following relationships hold for [J]:

$$[Jb]^t \; = \; \left[\begin{array}{c|c} A & 0 \\ \hline RA & A \end{array} \right] \qquad\qquad [Jb]^{-1} \; = \; \left[\begin{array}{c|c} A & RA \\ \hline 0 & A \end{array} \right]$$

$$[Jb]^{-t} \qquad = \qquad \left[\begin{array}{c|c} A^t & 0 \\ \hline A^t R^t & A^t \end{array} \right]$$

A.2 Matrix Method for Under Determined Finger Motions

For the case in which the motion of the object does not completely determine the motion of the finger, the potential energy may be minimized subject to the nc constraint conditions in [P]. The constraint equations, C_i, are formed by multiplying one row of [P] by d_c. The nf auxiliary equations may then be written as [139]

$$\varphi_i = \frac{\partial P.E.}{\partial q_i} + \lambda_1 \frac{\partial C_1}{\partial q_i} + \lambda_2 \frac{\partial C_2}{\partial q_i} + \cdots + \lambda_{nc} \frac{\partial C_{nc}}{\partial q_i}$$

These are combined with the constraint equations to provide $nf + nc$ equations for $nf + nc$ unknowns. The equations may be conveniently expressed as

$$\left[\begin{array}{c} d_q \\ \hline l \end{array} \right] \; = [L]^{-1} \left[\begin{array}{c} g_q \\ \hline d_c \end{array} \right]$$

where l is a column vector of the nc Lagrange multipliers and

$$[L] \; = \; \left[\begin{array}{c|c} Kq & P^t \\ \hline P & 0 \end{array} \right]$$

A.3 Differential Jacobians

In Section 5.3.1 the change in the jacobians, $[J]$, as a result of small displacements of the object are considered. These terms, $\Delta[J]$ and $\Delta[J]^t$, result from shifting of the contact area and rolling of the fingers. Products such as $\Delta[J]\cdot d$ contain very small terms and may be ignored, but products such as $\Delta[J]^t\cdot g$ may contain significant terms since the forces, g, may be large. As an example, if the contact area translates and rotates with respect to the object then the change in the jacobian relating g_{bp} and g_b is

$$\Delta[Jb]^t = [Jb']^t - [Jb]^t$$

where $[Jb']^t$ is the jacobian relating to the new position and orientation of the contact area and $[Jb]^t$ is the original jacobian. By writing $[Jb']^t$ and $[Jb]^t$ in terms of partitions, $\Delta[Jb]^t$ is seen to be

$$\begin{bmatrix} \Delta A & | & 0 \\ \hline \Delta(RA) & | & \Delta A \end{bmatrix}$$

where $\Delta(RA) = (RA)' - (RA) = [R][A] + [\Delta R][A] + [R][\Delta A] + [\Delta R][\Delta A] - [R][A]$. $[\Delta R][\Delta A]$ contains second order terms, and may be dropped so that $\Delta(RA) \approx [\Delta R][A] + [R][\Delta A]$.

$[\Delta R]$ and $[\Delta A]$ can be written in terms of differential translations and rotations, $(\delta r_x, \delta r_y, \delta r_z, \delta\theta_x, \delta\theta_y, \delta\theta_z)$.

$$[\Delta R] = [R'] - [R] = \begin{bmatrix} 0 & -\delta r_z & \delta r_y \\ \delta r_z & 0 & -\delta r_x \\ -\delta r_y & \delta r_x & 0 \end{bmatrix}$$

$$[\Delta A] = [A'] - [A] = \begin{bmatrix} 0 & -\delta\theta_z & \delta\theta_y \\ \delta\theta_z & 0 & -\delta\theta_x \\ -\delta\theta_y & \delta\theta_x & 0 \end{bmatrix}$$

$[\Delta A]$ and δr are also equivalent to the upper left 3x3 partition and right column respectively of the differential 4x4 homogeneous transform, $[\Delta]$, expressing a small translation and rotation of one coordinate system with respect to another [113].

A.4 Rolling Contact

$r'_{(s)} = r_{(s + \delta s)}$ and $u'_{(s)} = u_{(s + \delta s)}$ may be expanded in terms of $r_{(s)}$ as

$$r'_{(s)} = r_{(s)} + \delta s \frac{dr_b}{ds} + \frac{(\delta s)^2}{2!} \frac{d^2 r_b}{ds^2} + \cdots$$

$$u'_{(s)} = \frac{dr}{ds} + \delta s \frac{d^2 r}{ds^2} + \cdots$$

Then Δr becomes

$$\Delta r = \delta s \frac{dr}{ds} + \frac{(\delta s)^2}{2!} \frac{d^2 r}{ds^2} + \cdots = \delta s u + \frac{(\delta s)^2}{2!} \frac{du}{ds} + \cdots$$

Since the first derivatives of r_b and r_f are equal at the initial contact point, subtracting $\Delta r_b - \Delta r_f$ gives

$$\Delta r_b - \Delta r_f = \frac{(\delta s)^2}{2!} \left(\frac{d^2 r_b}{ds^2} - \frac{d^2 r_f}{ds^2} \right) + \cdots$$

or, $\Delta r_b - \Delta r_f \approx 1/2 (\delta s)^2$ times the difference in curvature between $r_{b(s)}$ and $r_{f(s)}$.

The rotation of the contact point is given by the vector $(u_b \times u_b')$ and the rotation of the fingertip is given by $(u_f' \times u_b')$. Expanding u_f' and u_b' in terms of $r_{(s)}$ and discarding third and higher derivatives of r gives

$$u_b \times u_b' = u_b \times (u_b + \frac{du_b}{ds} \delta s) = (0) + \delta s (\frac{dr_b}{ds} \times \frac{d^2 r_b}{ds^2})$$

and

$$u_f' \times u_b' = (u_f \times u_b) + \delta s (u_f \times \frac{du_b}{ds}) + \delta s (\frac{du_f}{ds} \times u_b)$$
$$+ (\delta s)^2 (\frac{du_f}{ds} \times \frac{du_b}{ds})$$

$$= (0) + \delta s \left((\frac{d^2 r_f}{ds^2} - \frac{d^2 r_b}{ds^2}) \times u \right) + (\delta s)^2 (\frac{d^2 r_f}{ds^2} \times \frac{d^2 r_b}{ds^2})$$

where $u = u_b = u_f$ at the initial contact point.

For the example in Section 5.5.3, where the object surface is flat and the fingertip is a segment of a circular arc, as in Figures 5-21 and 5-22, the rolling equations become

$$r_f = (r_f \sin \theta_f) i - (r_f \cos \theta_f) j, \quad r_b = (\tfrac{w}{2} \tan \theta_b) i + \tfrac{w}{2} j.$$

where θ_f is related to s as

$$\frac{d\theta_f}{ds} = 1/r_f$$

For $\theta_f = \theta_b = 0$ at the initial contact point, equations (5.14)-(5.17) become

$$\Delta r_b = (r_f \delta\theta_f) i + 0j$$

$$\Delta r_b - \Delta r_f = 0i - \frac{r_f(\delta\theta_f)^2}{2} j \approx 0$$

$$u_b \times u_b' = 0$$

$$u_f' \times u_b' = -\delta\theta_f k$$

A.5 Details for Examples in Section 5.5

Summary of matrix equations for left finger — point-contact example

1. $d_{bp} = [M][Jb]d_b$

2. $d_{fp} = [Jfq]d_q$

3. .
$$\begin{bmatrix} d_q \\ \hline \lambda_1 \\ \lambda_2 \end{bmatrix} = [L]^{-1} \begin{bmatrix} g_q \\ \hline dm \\ dn \end{bmatrix}$$

4. $\delta g_q = [Kq]d_q$

5. $[Cfp] = [Jfq][Kq]^{-1}[Jfq]^t$

6. $[Cfc]$ = non-singular portion of $[Cfp]$

7. d_{fc} = subset of d_{fp} corresponding to $[Cfc]$

8. $\delta g_{fc} = [Cfc]^{-1}d_{fc}$

9. $\delta g_{fp} = \delta g_{fc} + \Delta[Jf]^{-t}g_f$

10. $\delta g_b = [Jb]^t[M]^t \delta g_{fp}$

$$[Jb] = \begin{bmatrix} 0 & 0 & 1 & | & 0 & \frac{w}{2} & 0 \\ 0 & 1 & 0 & | & 0 & 0 & -\frac{w}{2} \\ -1 & 0 & 0 & | & 0 & 0 & 0 \\ \hline 0 & 0 & 0 & | & 0 & 0 & 1 \\ 0 & 0 & 0 & | & 0 & 1 & 0 \\ 0 & 0 & 0 & | & -1 & 0 & 0 \end{bmatrix}$$

$$[Jf] = \begin{bmatrix} 0 & 0 & 1 & | & 0 & r_f & 0 \\ 0 & 1 & 0 & | & 0 & 0 & r_f \\ -1 & 0 & 0 & | & 0 & 0 & 0 \\ \hline 0 & 0 & 0 & | & 0 & 0 & 1 \\ 0 & 0 & 0 & | & 0 & 1 & 0 \\ 0 & 0 & 0 & | & -1 & 0 & 0 \end{bmatrix}$$

$$[Jq] = \begin{bmatrix} 1 & 0 & 0 \\ 0 & 1 & 0 \\ 0 & 0 & 0 \\ \hline 0 & 0 & 0 \\ 0 & 0 & 0 \\ 0 & 0 & 1 \end{bmatrix}$$

$$[Jfq] = \begin{bmatrix} 0 & 0 & 0 \\ \hline 0 & 1 & r_f \\ -1 & 0 & 0 \\ \hline 0 & 0 & 1 \\ 0 & 0 & 0 \\ 0 & 0 & 0 \end{bmatrix} = [P]$$

$$[M] = \begin{bmatrix} 1 & 0 & 0 & | & 0 & 0 & 0 \\ 0 & 1 & 0 & | & 0 & 0 & 0 \\ 0 & 0 & 1 & | & 0 & 0 & 0 \end{bmatrix}$$

$$[Kq] = \begin{bmatrix} k_a & 0 & 0 \\ 0 & k_b & 0 \\ 0 & 0 & k_c \end{bmatrix}$$

$$[L] = \begin{bmatrix} k_a & 0 & 0 & | & 0 & -1 \\ 0 & k_b & 0 & | & 1 & 0 \\ 0 & 0 & k_c & | & r_f & 0 \\ \hline 0 & 1 & r_f & | & 0 & 0 \\ -1 & 0 & 0 & | & 0 & 0 \end{bmatrix}$$

$$[Cfp] = \begin{bmatrix} 0 & | & 0 & 0 & | & 0 & 0 \\ \hline 0 & | & \dfrac{k_b r_f^2 + k_c}{k_b k_c} & 0 & \dfrac{r_f}{k_c} & | & 0 & 0 \\ 0 & | & 0 & \dfrac{1}{k_a} & 0 & | & 0 & 0 \\ 0 & | & \dfrac{r_f}{k_c} & 0 & \dfrac{1}{k_c} & | & 0 & 0 \\ \hline 0 & | & 0 & 0 & 0 & | & 0 & 0 \\ 0 & | & 0 & 0 & 0 & | & 0 & 0 \end{bmatrix}$$

$$[Kfc] = \begin{bmatrix} k_b & 0 & -k_b r_f \\ 0 & k_a & 0 \\ -k_b r_f & 0 & k_b r_f^2 + k_c \end{bmatrix}$$

$$\begin{bmatrix} d_q \\ \hline \lambda_1 \\ \lambda_2 \end{bmatrix} = [L]^{-1} \begin{bmatrix} g_q \\ dm \\ dn \end{bmatrix}$$

Figure A-1: Matrices for left finger - pointed or rolling contact

$\Delta[Jf]^{-t}$ point contact

$$\begin{bmatrix} 0 & 0 & 0 & | & 0 & 0 & 0 \\ -\delta\theta & 0 & 0 & | & 0 & 0 & 0 \\ 0 & -\delta\theta & 0 & | & 0 & 0 & 0 \\ \hline 0 & 0 & 0 & | & 0 & 0 & 0 \\ 0 & 0 & 0 & | & -\delta\theta & 0 & 0 \\ 0 & 0 & -\delta\theta\, r_f & | & 0 & -\delta\theta & 0 \end{bmatrix}$$

$\Delta[Jf]^{-t}$ rolling contact

$$\begin{bmatrix} 0 & 0 & 0 & | & 0 & 0 & 0 \\ -\delta\theta & 0 & 0 & | & 0 & 0 & 0 \\ 0 & -\delta\theta & 0 & | & 0 & 0 & 0 \\ \hline \delta\theta\, r_f & 0 & 0 & | & 0 & 0 & 0 \\ 0 & 0 & 0 & | & -\delta\theta & 0 & 0 \\ 0 & 0 & 0 & | & 0 & -\delta\theta & 0 \end{bmatrix}$$

$\Delta[Jb]^{t}$ rolling contact

$$\begin{bmatrix} 0 & 0 & 0 & | & 0 & 0 & 0 \\ 0 & 0 & 0 & | & 0 & 0 & 0 \\ 0 & 0 & 0 & | & 0 & 0 & 0 \\ \hline \delta\theta\, r_f & 0 & 0 & | & 0 & 0 & 0 \\ 0 & 0 & 0 & | & 0 & 0 & 0 \\ 0 & 0 & \delta\theta\, r_f & | & 0 & 0 & 0 \end{bmatrix}$$

	Object			Finger		
	d_b	d_{bp}	d_c	d_{fp}	d_f	d_q
1	dx	0	0	0	$-dn$	$-dn$
2	dy	dm	dm	dm	$\dfrac{\beta\,dm}{k_b}$	$\dfrac{\beta\,dm}{k_b}$
3	0	dn	dn	dn	0	
4	0	$d\theta_l$		αdm	0	
5	0	0		0	0	
6	$d\theta_z$	0		0	αdm	αdm

In the above,

$dn = -dx$ 　　 $\delta\theta = d\theta_z - \alpha(dy - \dfrac{w\ d\theta_z}{2})$ 　　 $d_{bp} = [Jb]\ d_b$

$dm = dy - \dfrac{w\ d\theta_z}{2}$ 　　 $\beta = \dfrac{k_b\, k_c}{k_b\, r_f^2 + k_c}$ 　　 $d_c = [M]\ d_{bp}$

$d\theta_l = d\theta_z$ 　　　　　　　　　　　　　　　　　 $d_{fp} = [Jf]\ d_f$

$\alpha = \dfrac{k_b\, r_f}{k_b\, r_f^2 + k_c}$ 　　 $d_f = [Jq]\ d_q$

Figure A-2: Matrices for left finger - pointed or rolling contact

$$[Cfp]$$

$$
\begin{bmatrix}
\dfrac{3}{k_p} & 0 & 0 & 0 & 0 & 0 \\[2ex]
0 & \dfrac{1}{\beta}+\dfrac{3}{k_p} & 0 & \dfrac{r_f}{k_c} & 0 & 0 \\[2ex]
0 & 0 & \dfrac{1}{k_b}+\dfrac{1}{k_p} & 0 & 0 & 0 \\[2ex]
0 & \dfrac{r_f}{k_c} & 0 & \dfrac{1}{Bk_p}+\dfrac{1}{k_c} & 0 & 0 \\[2ex]
0 & 0 & 0 & 0 & \dfrac{1}{Bk_p} & 0 \\[2ex]
0 & 0 & 0 & 0 & 0 & \dfrac{3}{2Bk_p}
\end{bmatrix}
$$

Combined compliance for finger and square fingertip where $k_a = k_b$ in finger joint coordinates and

$$k_{comp} = k_p$$

$$k_{shear} = \frac{k_p}{3}$$

$$k_{bend} = Bk_p$$

$$k_{twist} = \frac{2k_p}{3}$$

$$B = \frac{Area}{12}$$

for the elastic fingertip

Figure A-3: Compliance matrix for soft finger example

References

1. W. B. Gevarter, "An Overview of Artificial Intelligence and Robotics Volume II," PB82-204439 NBSIR 82-2479, National Bureau of Standards, March 1982.

2. J.M. Hollerbach, "A Recursive Langrangian Formulation of Manipulator Dynamics," MIT AI Memo 533, Massachusetts Institute of Technology , June 1980.

3. J.Y.S. Luh, M.W. Walker and R.P. Paul, "On-Line Computational Scheme for Mechanical Manipulators," *Journal of Dynamic Systems, Measurement and Control (ASME)*, June 1980, pp. 69-76.

4. J.Y.S. Luh and C.S. Lin, "Scheduling of Parallel Computation for a Computer-Controlled Mechanical Manipulator," *IEEE Transactions on Systems, Man and Cybernetics*, Vol. SMC-12, No. 2, March 1982, .

5. S. Megahed, "A New Lagrangian Formulation of Manipulator Dynamics," *11th International Symposium on Industrial Robots*, Tokyo, Japan, October 1981, pp. 765-771.

6. R. Featherstone, "Position and Velocity Transformations between Robot End Effector Coordinates and Joint Angles," DAI Working Paper 92, Department of Artificial Intelligence, University of Edinburgh, July 1981.

7. Anon., "High Accuracy Robot Sytem for Target Positioning," Tech. report TR-1004A, Contraves Goerz Corp., 1984.

8. A. Liegeios, E. Dombre and P. Borrel, " Learning and Control for a Compliant Computer Controlled Manipulator," *IEEE Transactions on Automatic Control*, Vol. AC-29, No. 6, December 1980, pp. 1097-1980.

9. R.H. Cannon and E. Schmitz, "Initial Experiments on the End-Point Control of a Flexible One Link Robot," Tech. report, Stanford University, Department of Aeronautics/Astronautics, November 1983.

10. W.J. Book and M. Majett, "Controller Design for Flexible, Distributed Parameter Mechanical Arms Via Combined State Space and Frequency Domain Techniques," *Robotics Research and Advanced Applications*, ASME, Phoenix, AZ, November 1982, pp. 101-120.

11. A. Zalucky and D.E. Hardt, "Active Control of Robot Structure Deflections," *Journal of Dynamic Systems, Measurement and Control*, Vol. 106, No. 1, March 1984, pp. 63-69.

12. A. Liegeois, A. Fournier and M.J. Aldon, "Model Reference Control of High-Velocity Industrial Robots," *Joint Automatic Control Conference*, San Francisco, CA, August 1980.

13. R. Horowitz and M. Tomizuka, "An Adaptive Control Scheme for Mechanical Manipulators- Compensation of Nonlinearity and Decoupling Control," *American*

Society of Mechanical Engineers Winter Annual Meeting, Chicago, ILL, November 1980.

14. C.S.G. Lee and B.H. Lee, "Resolved Motion Adaptive Control for Mechanical Manipulators," *ASME Journal of Dynamic Systems, Measurement and Control,* Vol. 106, No. 2, June 1984, pp. 134-142.

15. R.P. Anex and M. Hubbard, "Modeling and Adaptive Control of a Mechanical Manipulator," *ASME Journal of Dynamic Systems, Measurement and Control,* Vol. 106, No. 3, September 1984, pp. 211-217.

16. H. Asada and T. Kanade, "Design Concept of Direct Drive Manipulators Using Rare-Earth DC Torque Motors," *7th International Joint Conference on Artificial Intelligence,* Vancouver, BC, Canada, August 1981.

17. H. Asada and I.H. Ro, "A Linkage Design for Direct-Drive Robot Arms," Paper No. 84-DET-143, ASME, 1984.

18. M. Mason, "Compliance and Force Control for Computer Controlled Manipulators," in *Robot Motion Planning and Control,* M. Brady *et al,* ed., The MIT Press, Cambridge, MA, 1982, pp. 373-404, ch. 5.

19. K. Salisbury, "Active Stiffness Control of a Manipulator in Cartesian Coordinates," *Proc. 19th IEEE Conference on Decision and Control,* Albuquerque, NM, December 1980, pp. 87-97.

20. D.E. Whitney, "Force Feedback Control of Manipulator Fine Motions," *Journal of Dynamic Systems, Measurement and Control (ASME),* June 1977, pp. 91-97.

21. D.E. Whitney *et al,* "Servo Controlled Mobility Device," United States Patent No. 4,156,835, May,1979

22. R. Paul and B.E. Shimano, "Compliance and Control," *The 1976 Joint Automatic Controls Conference,* Albuquerque, NM, December 1976, pp. 694-699.

23. R.P.C. Paul and B. Shimano, "Compliance and Control," in *Robot Motion Planning and Control,* M. Brady *et al,* ed., The MIT Press, Cambridge, MA, 1982, pp. 404-418, ch. 5.

24. H. Inoue, "Force Feedback in Precise Assembly Tasks," MIT AI Memo 308, Massachusetts Institute of Technology, August 1974.

25. T. Goto, T. Onoyama and K. Takeyasu, "Precise Insert Operation by Tactile Controlled Robot HI-T-HAND Expert 2," *Proc. 4th International Symposium on Industrial Robots,* Tokyo, Japan, November 1974.

26. Tatsuo Goto, Kiyoo Takeyasu and Tadao Inoyama, "Control Algorithm for Precision Insert Operation Robots," *IEEE Transactions on Systems, Man and Cybernetics,* Vol. SMC-10, No. No. 1, Jan 1980, pp. 19-24.

27. G. Piller, "A Compact Six-Degree-of-Freedom Force Sensor For Assembly Robot,"

Proceedings, 12th International Symposium on Industrial Robots, Paris, France, June 1982, pp. 121-129.

28.   M.R. Cutkosky and P.K. Wright, "Position Sensing Wrists for Industrial Manipulators," *12th International Symposium on Industrial Robots,* Paris, France, June 1982.

29.   M.R. Cutkosky, J.M. Jourdain and P.K. Wright, "Testing and Control of Compliant Wrist," *14th International Symposium on Industrial Robots,* Gothenburg, Sweden, October 1984.

30.   H. Asada and H. Yamamoto, "Torque Feedback Control of MIT Direct-Drive Robot," *14th International Symposium on Industrial Robots,* Gothenburg, Sweden, October 1984, pp. 663-670.

31.   R. Paul, *Modelling, Trajectory Calculation and Servoing of a Computer Controlled Arm,* PhD thesis, Computer Science Department, Stanford University, August 1972.

32.   R. Paul, "Manipulator Path Control," *The 1975 International Conference on Cybernetics and Society,* , 1975, pp. 147-152.

33.   M.H. Raibert and J.J. Craig, "Hybrid Position/Force Control of Manipulators," *ASME Journal of Dynamic Systems, Measurement, and Control,* Vol. 102, June 1981, pp. 126-133.

34.   C.H. Wu and R.P. Paul, "Resolved Motion Force Control of Robot Manipulator," *IEEE Transactions on Systems, Man and Cybernetics,* Vol. SMC-12, No. 3, May 1982, pp. 266-275.

35.   H. Ozaki, A. Mohri and M. Takata, " On The Force Feedback Control of a Manipulator With a Compliant Wrist Force Sensor," *Mechanism and Machine Theory (GB),* Vol. 18, No. 1, 1983, pp. 57-62.

36.   N. Hogan and S.L. Cotter, "Cartesian Impedance Control of a Nonlinear Manipulator," *Robotics Research and Advanced Applications,* ASME Winter Annual Meeting, Phoenix, AZ, November 1982, pp. 121-128.

37.   N. Hogan, "Impedance Control of Industrial Robots," *Robotics and Computer-Inegrated Manufacturing,* Vol. 1, No. 1, 1984, pp. 97-113.

38.   D.E. Whitney *et al,* "Part Mating Theory for Compliant Parts," First Report, The Charles Stark Draper Laboratory Inc., August 1980, NSF Grant No. DAR79-10341.

39.   H. Van Brussel and J. Simons, "The Adaptable or Compliance Concept and its use for Automatic Assembly by Active Force Feedback Accomodations," *9th International Symposium on Industrial Robots,* Washington, DC, , 1979, pp. 167-181.

40.   H. Van Brussel, H. Thielmans and J. Simons, "Further Developments of the Active Adaptive Compliant Wrist (AACW) For Robot Assembly," *11th International Symposium on Industrial Robots,* Tokyo, Japan, October 1981, pp. 377-384.

41. A. Sharon and D. Hardt, "Enhancement of Robot Accuracy Using Endpoint Feedback and a Macro-Micro Manipulator System," Research Report RC 10440, I.B.M., March 1984.

42. J.L. Nevins et al, "Exploratory Research in Industrial Parts Mating," Seventh Progress Report for The National Science Foundation, The Charles Stark Draper Laboratory Inc., Cambridge, MA, February 1980.

43. S.H. Drake, P.C. Watson and S.H. Simunovic, "High Speed Assembly of Precision Parts Using Compliance Instead of Sensory Feedback," *Proc. 7th International Symposium on Industrial Robots,* Tokyo, Japan, , 1977, pp. 87-97.

44. H. McCallion, K.V. Alexander and D.T. Pham, "Aids for Automatic Assembly," *1st International Conference on Assembly Automation,* Brighton, UK, March 1980, pp. 87-97.

45. A. Zalucky and D.E. Hardt, "Active Control of Robot Structure Deflections," *Robotics Research and Advanced Applications,* ASME, Phoenix, AZ, November 1982, pp. 83-100.

46. A.C. Sanderson and G. Perry, "Sensor-Based Robotic Assembly Systems: Research and Applications in Electronic Manufacturing," *Proceedings of the IEEE,* Vol. 71, No. 7, July 1983, pp. 856-871.

47. G. Hirzinger, "Robots with Force-torque Sensing," *Process Automation,* 1982, pp. 8-12.

48. D.E. Whitney, "Quasi-Static Assembly of Compliantly Supported Parts," *ASME Transactions on Dynamic Systems, Measurement and Control,* Vol. 104, No. 1, March 1982, pp. 65-77.

49. D.E. Whitney, R.E. Gustavson and M.P. Hennessey, "Designing Chamfers," *The International Journal of Robotics Research,* Vol. 2, No. 4, 1983, pp. 3-18.

50. M.S. Ohwovorioli, *An Extension of Screw Theory and its Application to the Automation of Industrial Assemblies,* PhD thesis, Stanford University, April 1980.

51. C.C. Selvage, "Assembly of Interference Fits by Impact and Constant Force Methods," Master's thesis, Massachusetts Institute of Technology, June 1979.

52. F.W. Paul, T.K. Gettys and J.D. Thomas, "Definning of Iron Castings Using a Robotic Positioned Chipper," *Robotics Research and Advanced Applications,* W.J. Book, ed., ASME, Phoenix, AZ, November 1982, pp. 269-278.

53. E. Abele, D. Boley and W. Sturz, "Interactive Programming of Industrial Robots for Deburring," *14th International Symposium on Industrial Robots,* Gothenburg, Sweden, October 1984, pp. 505-515.

54. H. Asada and Y. Sawada, "Design of an Adaptable Tool Guide for Grinding Robots," Paper No. 84-DET-41, ASME, 1984.

55. P.R. Fitzpatrick and J.J. Barto, "Automated Drilling and Riveting System," *Applying Robotics in the Aerospace Industry*, SME, March 1984, pp. MS84-218, 1-15.

56. T.F. Gayou and T.F. Hogan, "A Graphical Robot Programming System for Composite Routing," *Applying Robotics in the Aerospace Industry*, SME, March 1984, pp. MS84-215, 1-11.

57. W. Marx, "Robotic Drilling and Routing for Complex Parts," *Applying Robotics in the Aerospace Industry*, SME, March 1984, pp. MS84-223, 1-11.

58. G. Lundstrom, B. Glemme, B.W. Rooks, *Industrial Robots - Gripper Review*, International Fluidics Services Ltd., 35-39 High Street, Kempston, Bedford, England, 1977.

59. I. Kato, *Mechanical Hands Illustrated*, Hemisphere Publishing Corporation, New York, N.Y., 1982.

60. P.K. Wright and M.R. Cutkosky, "Design of Grippers," in *Handbook of Industrial Robotics*, S. Nof, ed., John Wiley and Sons, Inc., New York, NY, 1984, ch. 2.4.

61. M.R. Cutkosky and E. Kurokawa, "Grippers for an Unmanned Forging Cell," Robotics Institute Technical Report CMU-RI-TR-83-3, Carnegie-Mellon University, April 1983.

62. S. Hirose and Y. Umetani, "The Development of Soft Gripper for the Versatile Robot Hand," *Mechanism and Machine Theory (GB)*, 1978, pp. 351-358.

63. A. Rovetta, "On Specific Problems of Design of Multipurpose Mechanical Hands in Industrial Robots," *Seventh International Symposium on Industrial Robots*, Tokyo, Japan, , 1977, pp. 337-343.

64. Anon, "Inside Japan," *Assembly Automation (GB)*, February 1982, pp. 58.

65. H.J. Warnecke and I.I. Schmidt, " Flexible Grippers For Handling Systems - Design Possibilities and Experiences," *5th I.C.P.R.*, Chicago, IL, August 1979, pp. 320-324.

66. G. Bancon and B. Huber, " Depression and Dual Grippers With Their Possible Applications," *Proceedings, 12th International Symposium on Industrial Robots*, Paris, France, June 1982, pp. 321-329.

67. R. Tella, J.R. Birk and R.B. Kelley, "General Purpose Hands for Bin-Picking Robots," *IEEE Transactions on Systems, Man and Cybernetics*, Vol. SMC-12, No. 6, Nov 1982, pp. 828-837.

68. G. Lundstrom, "A New Method of Designing Grippers," *Sixth International Symposium on Industrial Robots*, Nottingham, UK, March 1976, pp. 261-268.

69. F. Skinner, "Designing a Multiple Prehension Manipulator," *Journal of Mechanical Engineering*, Vol. 97, No. 9, September 1975, pp. 30-37.

70. N.G. Caruso, "Robotic Assembly: Three Fingered Gripper," *Applying Robotics in the Aerospace Industry*, SME, March 1984, pp. MS84-228, 1-18.

71.　Anon., " Cranfield Offers a Unique Programme for Training Engineers," *Assembly Automation (UK)*, November 1982, pp. 194-196.

72.　F.Y. Chen, "Force Analysis and Design Considerations of Grippers," *Industrial Robot (UK)*, Vol. 9, No. 4, December 1982, pp. 243-249.

73.　I.B. Chelpanov and S.N. Kolpashnikov, "Problems With The Mechanics of Industrial Robot Grippers," *Mechanism and Machine Theory*, Vol. 18, No. 4, 1983, pp. 295-299.

74.　D.D. Grossman and M.W. Blasgen, "Orienting Mechanical Parts by Computer Controlled Manipulator," *IEEE Transactions on Systems, Man and Cybernetics*, Vol. SMC-5, No. 9, September 1975, pp. 561-565.

75.　T. Okada, "Object Handling System for Manual Industry," *IEEE Transactions on Systems, Man, and Cybernetics*, Vol. SMC-9, No. 2, February 1979, pp. 79-89.

76.　T. Okada, "Computer Control of Multijointed Finger System for Precise Handling," *IEEE Transactions on Systems, Man and Cybernetics*, Vol. SMC-12, No. 3, May 1982, pp. 289-299.

77.　J.K. Salisbury and J.J. Craig, "Articulated Hands: Force Control and Kinematic Issues," *Robotics Research*, Vol. 1, No. 1, 1982, pp. 4-17.

78.　J.K. Salisbury, *Kinematic and Force Analysis of Articulated Hands*, PhD thesis, Stanford University, July 1982.

79.　J.D. Abramowitz, J.W. Goodnow and B. Paul, "Pennsylvania Articulated Mechanical Hand," *International Conference on Computers in Engineering*, ASME, Chicago, IL, August 1983.

80.　Y. Nakano, M. Fujie and Y. Hosada, "Hitachi's Robot Hand," *Robotics Age*, Vol. 6, No. 7, July 1984, pp. 18-20.

81.　S.C. Jacobsen, J.E. Wood, D.F. Knutti and K.B. Biggers, "The Utah/MIT Dextrous Hand: Work in Progress," *First International Conference on Robotics Research*, M. Brady and R.P. Paul, ed., MIT Press, Cambridge, Mass., , 1984, pp. 601-653.

82.　J.R. Birk, "A Computer Controlled Rotating-Belt Hand for Object Orientation," *IEEE Transactions on Systems, Man and Cybernetics*, Vol. SMC-4, No. 2, March 1974, pp. 186-191.

83.　P. Datseris and W. Palm, "Principles on the Development of Mechanical Hands Which Can Manipulate Objects by Means of Active Control," Paper 84-DET-37, ASME, 1984.

84.　J.M. Hollerbach, "Workshop on the Design and Control of Dexterous Hands," A.I. Memo 661, Massachusetts Institute of Technology, April 1982.

85.　H. Asada, *Studies on Prehension and Handling by Robot Hands With Elastic Fingers*, PhD thesis, Kyoto University, April 1979.

86. H. Hanafusa, K. Kobayashi and N. Terasaki, "Fine Control of the Object With Articulated Multi-Finger Robot Hands," *1983 International Conference on Advanced Robotics,* Tokyo, Japan, September 1983, pp. 245-252.

87. D.E. Orin and S.Y. Oh, "Control of Force Distribution in Robotic Mechanisms Containing Closed Kinematic Chains," *Journal of Dynamic Systems, Measurement and Control (ASME),* Vol. 102, June 1981, pp. 134-141.

88. M.T. Mason, *Manipulator Grasping and Pushing Operations,* PhD thesis, Massachusetts Institute of Technology, June 1982.

89. R. Malek, "The Grip and its Modalities," in *The Hand,* R. Tubiana, ed., W.B. Saunders Co., Philadelphia, PA, 1981, ch. 45.

90. R.S. Fearing, "Simplified Grasping and Manipulation With Dextrous Robot Hands," *1984 American Control Conference,* San Diego, CA, June 1984, pp. 32-38.

91. J.D. Wolter, R.A. Volz and A.C. Woo, "Automatic Generation of Gripping Positions," Tech. report RSD-TR-2-84, Center for Robotics and Integrated Manufacturing, The University of Michigan, February 1984.

92. R. Tomovic and G. Boni, "An Adaptive Artificial Hand," *IRE Transactions on Automatic Control,* Vol. AC-7, No. 3, April 1962, pp. 3-10.

93. K. Corker, A. Mishkin and J. Lyman, "Achievement of a Sense of Operator Presence in Remote Manipulation," Biotechnology Laboratory Technical Report 60, University of California, Los Angeles, October 1980.

94. J.C. Baits, R.W. Todd and J.M. Nightingale, " The Feasibility of an Adaptive Control Scheme for Artificial Prehension," *Proceedings of the Institution of Mechanical Engineers (GB),* Vol. 183, No. Pt. 3-F, 1968-69, pp. 54-59.

95. N.D. Ring and D.B. Welbourn, "A Self-Adaptive Gripping Device: It's Design and Performance," *Proceedings of the Institution of Mechanical Engineers (GB),* Vol. 183, No. Pt. 3-F, 1968-69, pp. 45-49.

96. E. Heer and A.K. Bejczy, " Control of Robot Manipulators for Handling and Assembly in Space," *Mechanism and Machine Theory (GB),* Vol. 18, No. 1, 1983, pp. 23-35.

97. M.R. Cutkosky, P.S. Fussell and R. Milligan Jr., "The Design of a Flexible Machining Cell for Small Batch Production," *Journal of Manufacturing Systems,* Vol. 3, No. 1, 1984, .

98. T. O. Binford *et al,* "Exploratory Study of Computer Integrated Assembly Systems," Fifth Report, Memo AIM-285.5, Artificial Intelligence Laboratory, Stanford University, September 1978.

99. J. Tlusty and M. Elbestawi, "Constraints in Adaptive Control with Flexible End Mills," *Annals of the CIRP,* Vol. 28, No. 1, 1979, pp. 253-255.

100. A.G. Ulsoy, Y. Koren and F. Rasmussen, "Principal Developments in the Adaptive Control of Machine Tools," *(ASME) Journal of Dynamic Systems, Measurement and Control,* Vol. 105, June 1983, pp. 107-112.

101. D.W. Yen and P.K. Wright, "Adaptive Control in Machining - A New Approach Based on the Physical Constraints of Tool Wear Mechanisms," *ASME Journal of Engineering for Industry,* Vol. 105, No. 1, February 1983, pp. 31-38.

102. D.W. Yen, *A New Approach and a Remote Temperature Sensing Technique for Adaptive Control Machining,* PhD thesis, Carnegie-Mellon University, 1984.

103. M.R. Cutkosky and P.K. Wright, "Compliance System for Industrial Manipulators," United States Patent No. 4,458,424, July 1984.

104. D.E. Whitney and E.F. Junkel, "Applying Stochastic Control Theory to Robot Sensing, Teaching and Long Term Control," *12th International Symposium on Industrial Robots,* Paris, France, June 1982.

105. A. Gelb *et al, Applied Optimal Estimation,* The M.I.T. Press, Cambridge, MA, 1974.

106. G.F. Franklin and J.D. Powell, *Digital Control of Dynamic Systems,* Addison-Wesley Inc., Reading, MA, 1980.

107. M. Shwarz and L. Shaw, *Signal Processing: Discrete Spectral Analysis, Detection and Estimation,* McGraw-Hill Inc., New York, NY, 1975.

108. A.E. Hunt and A.C. Sanderson, "Vision-Based Predictive Robotic Tracking of a Moving Target," Robotics Institute Technical Report CMU-RI-TR-82-15, Carnegie-Mellon University, 1982.

109. M. Davis, private communication, Manager, Manufacturing Development, Cincinnati Milacron Inc.

110. T. Lozano-Perez, "Task Planning," in *Robot Motion Planning and Control,* M. Brady *et al,* ed., The M.I.T. Press, Cambridge, MA, 1982, pp. 463-488, ch. 6.

111. T. Lozano-Perez, "Automatic Planning of Manipulator Transfer Movements," in *Robot Motion Planning and Control,* M. Brady *et al,* ed., The M.I.T. Press, Cambridge, MA, 1982, pp. 489-526, ch. 6.

112. B.E. Shimano, *The Kinematic Design and Force Control of Computer Controlled Manipulators,* PhD thesis, Stanford University, March 1978.

113. R.P. Paul, *Robot Manipulators: Mathematics Programming and Control,* The MIT Press, Cambridge, MA, 1981.

114. S.H. Crandall, N.C. Dahl and T.J. Lardner, *Theory of Elasticity,* McGraw-Hill, Inc., New York, 1972.

115. S.P. Timoshenko and J.N. Goodier, *Theory of Elasticity,* McGraw-Hill, Inc., New York, 1970.

116.   J. Halling, *Principles of Tribology*, Macmillan Press, Ltd., London, 1975.

117.   K.C. Ludema, "Friction of Rubber," in *Mechanics of Pneumatic Tires, NBS Monograph 122*, S.K. Clark, ed., U.S. Govt. Printing Office, Washington, D.C., 1971, pp. 41-54, ch. 1.2.

118.   J.Y. Wong, *Theory of Ground Vehicles*, John Wiley and Sons, Inc., New York, 1978.

119.   E. Grandjean, *Fitting the Task to the Man*, Taylor and Francis Ltd., London, 1981.

120.   J. Napier, "The Evolution of the Hand," *Scientific American*, Vol. 207, No. 6, December 1962, pp. 56-62.

121.   M. Cartmill, "Pads and Claws in Arboreal Locomotion," in *Primate Locomotion*, F.A. Jenkins, Jr., ed., Academic Press, New York, 1974, pp. 45-76, ch. 2.

122.   M.R. Cutkosky and P.K. Wright, " Grip Stability and the Design of Robotic Fingers," Robotics Institute Technical Report, Carnegie-Mellon University, i<tobepublishedin>1985.

123.   P. Rabischong, "Phylogeny of the Hand," in *The Hand*, R. Tubiana, ed., W.B. Saunders Co., Philadelphia, PA, 1981, ch. 4.

124.   R.S. Fearing, "Exploration of the Dextrous Hand Control Problem," Tech. report 82CRD337, General Electric Corporate Research and Development, December 1982.

125.   M.L. Nagurka, "Prosthetic Hand Devices," Tech. report, College of Engineering and Applied Science, University of Pennsylvania, March 1976.

126.   L. Bunnell, *Bunnell's Surgery of the Hand*, J.B. Lippincott Co., Philadelphia, PA, 1970.

127.   J. Imbriglia, M.D., private communication.

128.   G. Schlesinger, *Der Mechanische Aufbau der kunstlischen Glieder, Part II of Ersatzglieder und Arbeitshilfen*, Springer, Berlin, 1919.

129.   C.L. Tayor and R.J. Schwarz, "The Anatomy and Mechanics of the Human Hand," *Artificial Limbs*, Vol. 2, May 1955, pp. 22-35.

130.   S. Jacobson, "Neurocytology," in *An Introduction to the Neurosciences*, B.A. Curtis, S. Jacobson and E.M. Marcus, ed., W.B. Saunders Co., Philadelphia, 1972, pp. 36-71, ch. 3.

131.   E.E. Selkurt, *Basic Physiology for the Health Sciences*, Little, Brown and Co., Boston, 1975.

132.   B. Katz, *Nerve, Muscle and Synapse*, Mcgraw-Hill, Inc., New York, 1966.

133.   P.A. Merton, "How We Control the Contraction of Our Muscles," *Scientific American*, Vol. 226, No. 5, May 1972, pp. 30-37.

134. T.E. Twitchell, "The Automatic Grasping Responses of Infants," *Neuropsychologia,* Vol. 3, 1965, pp. 247-259.

135. E.M. Marcus, "The Motor System and the Integration of Reflex Activity," in *An Introduction to the Neurosciences,* B.A. Curtis, S. Jacobson and E.M. Marcus, ed., W.B. Saunders Co., Philadelphia, PA, 1972, ch. 23.

136. R. Ayres, N. Capell and S. Miller, "Experimental Results on Assembly Performance VS Amount of Sensory Information," Tech. report, Carnegie-Mellon University, September 1983.

137. E. Glassman, R.L. Hollis, Jr. and R.H. Taylor, "Compliant Gripper," *I.B.M. Technical Disclosure Bulletin,* Vol. 27, No. 1B, June 1984, pp. 772-773.

138. L.D. Harmon, "Robotic Taction for Industrial Assembly," *The International Journal of Robotics Research,* Vol. 3, No. 1, 1984, pp. 73-76.

139. S.B. Hildebrand, *Advanced Calculus for Applications,* Prentice-Hall, Inc., Englewood Cliffs, N.J., 1976.

Index

accuracy, 2, 5, 29, 36, 47, 156
actuators, 141, 156
adaptation, 25
adaptive control, 29
adhesion, 16, 95, 98, 104
animals, 1, 136
architecture, 37
area of contact, 119-121, 149, 154
   155
arms, 5
assembly, 10, 18-22, 32, 39, 153

bending, 99, 116

C-surface, 6, 7, 22, 26
cells, metal working, 2, 17, 24, 32
compliance, 6-8, 54, 105, 120, 142, 155
   center of, 20
compression, 101
constraint, 6, 22, 26, 59, 78, 79
contact 13, 56, 87; *see also* point
   contact, curved contact, soft
   contact
contact area, 119-121, 149, 154, 155
control,
   adaptive, 29
   architecture, 37, 152, 153
   arm, 5, 38, 46, 47
   force, 7, 30, 56, 150
   hand, 9, 150-152
   position, 29, 150
   wrist, 38, 122, 132
contour following, 26, 40
contour modification, 28
coordinate systems, 75
Coulomb friction, 15, 16, 58, 64, 95,
   154
curvature, 92, 94, 119-122

curved contact, 83, 91, 105, 113-116,
   154

deburring, 11, 52
deformation, 103, 120, 154
dynamics, 6, 19

elasticity, 59, 99, 136
end effector, 3
estimator, 42-45

filter, 42-45, 154
fine manipulation, 4, 5, 8, 30, 140,
   143-145, 152, 153, 156
fingers, 12, 55, 58, 59, 63, 76, 148
fingertips, 56, 87, 90, 92-107, 149
flat contact, 84, 94, 119
force control, 30, 56, 150
forging, 1
frequency, 25
friction, 10, 13, 15, 16, 56, 58, 64, 119,
   120, 124, 143, 154

grasping, 1, 3, 12-16, 23, 56, 126-131,
   143-145, 156
grinding, 11, 24, 25, 28, 29, 32,
   51-53, 129, 153
grip, 56, 61, 74, 86, 122
grippers, 1-3, 11, 57, 63

hands,
   control of, 5
   human, 3, 55, 123-137
   primate, 138, 139
   robotic, 12-14, 16, 154
hierarchy, 4, 150

impedance, 7, 61

industrial robot, 1, 5, 17, 18, 22

jacobian, 14, 76, 83, 86, 157-159
joint variables, 76

Kalman filter, 42-45

machine tools, 2, 28-29
machining, 29
machinist, 29, 126, 130
manipulation, *see* fine manipulation
manipulator, 1, 5, 8, 17, 148; *see also*
    industrial robot
manufacturing, 2
materials handling, 17
microcomputer, 32, 38
mobility, 78
muscles, 125, 128, 136, 145

noise, 27, 42
normal stress, 98

painting, 24, 25
point contact, 13, 90, 109, 113,
    119, 154
polishing, 24, 25
position control, 29, 150
potential energy, 81, 158
predictor, 43, 47-50
pressure, 106
primates, 138

radius of curvature, 94, 119, 121, 122
    154
reflexes, 133-135, 150, 152, 154
remote-center-compliance (RCC), 8,
    23, 30, 33, 140
rolling, 14, 92-94, 99, 107, 119, 154,
    160, 161
routing, 25, 28
rubber, 103, 105, 116

sanding, 24
sensors, 2, 32, 36, 55, 122, 132-134,
    137, 141, 148-150, 153, 156
shear, 100, 107
slipping, 16, 61, 100-101, 155
skin, 16, 96
soft contact, 95-99, 116-118, 154
soft, curved contact, 105-109
spheres, hydraulic, 32
stability, 13, 16, 56, 61, 69-71
    115, 119, 154
stiffness,
    arm, 5, 7
    finger, 63, 64, 76, 154
    fingertip, 102, 103, 116
    grip, 56, 61, 74, 122, 129, 150
    task, 7, 25, 150
    wrist, 33, 54, 155
surface finishing, 11, 23-26, 153

time dependence, 24, 29, 41
tools, 3
torsion, 101, 116
tracking, 26, 40, 41
transformation, 76-77, 157

welding, 24
worst-case finger, 65
wrists,
    human, 8, 125
    robotic, 8, 9, 22, 32, 141-143
    145, 146, 153